Trusted Computing Platforms:
Design and Applications

TRUSTED COMPUTING PLATFORMS: DESIGN AND APPLICATIONS

SEAN W. SMITH
Department of Computer Science
Dartmouth College
Hanover, New Hampshire USA

 Springer

Sean W. Smith
Dartmouth College
Department of Computer Science
Hanover, NH 03755
USA

Trusted Computing Platforms: Design and Applications

Library of Congress Cataloging-in-Publication Data

A C.I.P. Catalogue record for this book is available
from the Library of Congress.

ISBN 0-387-23916-2   e-ISBN 0-387-23917-0   Printed on acid-free paper.

Printed in the United States of America.

9 8 7 6 5 4 3 2 1   SPIN 11311096

springeronline.com

Contents

List of Figures

List of Tables

Preface

We stand an exciting time in computer science. The long history of specialized research building and using security-enhanced hardware is now merging with mainstream computing platforms; what happens next is not certain but is bound to be interesting. This book tries to provide a roadmap.

A fundamental aspect of the current and emerging information infrastructure is distribution: multiple parties participate in this computation, and each may have different interests and motivations. Examining security in these distributed settings thus requires examining which platform is doing what computation—and which platforms a party must trust, to provide certain properties despite certain types of adversarial action, if that party is to have trust in overall computation. Securing distributed computation thus requires considering the trustworthiness of individual platforms, from the differing points of view of the different parties involved. We must also consider whether the various parties in fact trust this platform—and if they should, how it is that they know they should.

The foundation of computing is hardware: the actual platform—gates and wires—that stores and processes the bits. It is common practice to consider the standard computational resources—e.g., memory and CPU power—a platform can bring to a computational problem. In some settings, it is even common to think of how properties of the platform may contribute to more intangible overarching goals of a computation, such as fault tolerance. Eventually, we may start trying to change the building blocks–the fundamental hardware—in order to better suit the problem we are trying to solve.

Combining these two threads—-the importance of trustworthiness in these Byzantine distributed settings, with the hardware foundations of computing platforms—gives rise to a number of questions. What are the right trustworthiness properties we need for individual platforms? What approaches can we try in the hardware and higher-level architectures to achieve these properties? Can we usefully exploit these trustworthiness properties in computing platforms for broader application security?

With the current wave of commercial and academic trusted computing architectures, these questions are timely. However, with a much longer history of secure coprocessing, secure boot, and other experimentation, these questions are not completely new. In this book, we will examine this big picture. We look at the depth of the field: what a trusted computing platform might provide, how one might build one, and what one might be done with one afterward. However, we also look at the depth of history: how these ideas have evolved and played out over the years, over a number of different real platforms—and how this evolution continues today.

I was drawn to this topic in part because I had the chance to help do some of the work that shaped this field. Along the way, I've enjoyed the privilege of working with a number of excellent researchers. Some of the work in this book was reported earlier in my papers [SW99, SPW98, Smi02, Smi01, MSWM03, Smi03, Smi04], as documented in the "Further Reading" sections. Some of my other papers expand on related topics [DPSL99, DLP+ 01, SA98, SPWA99, JSM01, IS03b, SS01, IS03a, MSW+ 04, MSMW03, IS04b, IS04a].

Acknowledgments

Besides being a technical monograph, this book also represents a personal research journey stretching over a decade.

I am not sure how to begin acknowledging all the friends and colleagues who assisted with this journey. To start with: I am grateful to Doug Tygar and Bennet Yee, for planting these seeds during my time at CMU and continuing with friendship and suggestions since; to Gary Christoph and Vance Faber at Los Alamos, for encouraging this work during my time there; and to Elaine Palmer at IBM Watson, whose drive saw the defunct Citadel project turn into a thriving research and product development effort. Steve Weingart and Vernon Austel deserve particular thanks for their collaborations with security architecture and formal modeling, respectively. Thanks are also due to the rest of the Watson team, including Dave Baukus, Ran Canetti, Suresh Chari, Joan Dyer, Bob Gezelter, Juan Gonzalez, Michel Hack, Jeff Kravitz, Mark Lindemann, Joe McArthur, Dennis Nagel, Ron Perez, Pankaj Rohatgi, Dave Safford, and David Toll; to the 4758 development teams in Vimercate, Charlotte, Poughkeepsie, and Lexington; and to Mike Matyas.

Since I left IBM, this journey has been helped by fruitful discussions with many colleagues, including Denise Anthony, Charles Antonelli, Dmitri Asonov, Dan Boneh, Ryan Cathecart, Dave Challener, Srini Devadas, John Erickson, Ed Feustel, Chris Hawblitzel, Peter Honeyman, Cynthia Irvine, Nao Itoi, Ruby Lee, Neal McBurnett, Dave Nicol, Adrian Perrig, Dawn Song, and Leendert van Doorn. In academia, research requires buying equipment and plane tickets and paying students; these tasks were supported in part by the Mellon Foundation, the NSF (CCR-0209144), AT&T/Internet2 and the Office for Domestic Preparedness, Department of Homeland Security (2000-DT-CX-K001).

Here at Dartmouth, the journey continued with the research efforts of students including Alex Barsamian, Mike Engle, Meredith Frost, Alex Iliev, Shan Jiang, Evan Knop, Rich MacDonald, John Marchesini, Kazuhiro Minami, Mindy Periera, Eric Smith, Josh Stabiner, Omen Wild, and Ling Yan. My colleagues in the Dartmouth PKI Lab and the Department of Computer Science also provided invaluable helpful discussion, and coffee too.

Dartmouth students Meredith Frost, Alex Iliev, John Marchesini, and Scout Sinclair provided even more assistance by reading and commenting on early versions of this manuscript.

Finally, I am grateful for the support and continual patience of my family.

Sean Smith
Hanover, New Hampshire
October 2004

Chapter 1

INTRODUCTION

Many scenarios in modern computing give rise to a common problem: why should Alice trust computation that's occurring at Bob's machine? (The computer security field likes to talk about "Alice" and "Bob" and protection against an "adversary" with certain abilities.) What if Bob, or someone who has access to his machine, is the adversary?

In recent years, industrial efforts—such as the *Trusted Computing Platform Association (TCPA)* (now reformed as the *Trusted Computing Group, TCG*), Microsoft's *Palladium* (now the *Next Generation Computing Base, NGSCB*), and Intel's *LaGrande*—have advanced the notion of a "trusted computing platform." Through a conspiracy of hardware and software magic, these platforms attempt to solve this remote trust problem, for various types of adversaries. Current discussions focus mostly on snapshots of the evolving TCPA/TCG specification, speculation about future designs, and idealogical opinions about potential social implications. However, these current efforts are just points on a larger continuum, which ranges from earlier work on *secure coprocessor* design and applications, through TCPA/TCG, to recent academic developments. Without wading through stacks of theses and research literature, the general computer science reader cannot see this big picture.

The goal of this book is to fill this gap. We will survey the long history of amplifying small amounts of hardware security into broader system security. We will start with early prototypes and proposed applications. We will examine the theory, design, implementation of the IBM 4758 secure coprocessor platform, and discuss real case study applications that exploit the unique capabilities of this platform. We will discuss how these foundations grow into the newer industrial designs such as TCPA/TCG, as well as alternate architectures this newer hardware can enable. We will then close with an examination of more recent cutting-edge experimental work.

1.1   Trust and Computing

We should probably first begin with some definitions. This book uses the term *trusted computing platform (TCP)* in its title and throughout the text, because that is the term the community has come to use for this family of devices.

This terminology is a bit unfortunate. "*Trusted* computing platform" implies that some party trusts the platform in question. This assertion says nothing about who that party is, whether the platform is worthy of that party's trust, and on what basis that party chooses to trust it. (Indeed, some wags describe "trusted computing" as computing which circumstances force one to trust, like it or not.)

In contrast, the devices we consider involve trust on several levels. The devices are, to some extent, *worthy of trust*: physical protections and other techniques protect them against at least some types of malicious actions by an adversary with direct physical access. A relying party, usually remote, has the ability to *choose to trust* that the computation on the device is authentic, and has not been subverted. Furthermore, typically, the relying party does not make this decision blindly; the device architecture provides some means to *communicate* its trustworthiness. (I like to use the term "trustable" for these latter two concepts.)

1.2   Instantiations

Many types of devices either fit this definition of "trusted computing platform," or have sufficient overlap that we must consider their contribution to the family's lineage.

We now survey the principal classes.

Secure Coprocessors.   Probably the purest example of a trusted computing platform is a *secure coprocessor*.

In computing systems, a generic *coprocessor* is a separate, subordinate unit that offloads certain types of tasks from the main processing unit. In PC-class systems, one often encounters *floating-point coprocessors* to speed mathematical computation. In contrast to these, a *secure coprocessor* is a separate processing unit that offloads security-sensitive computations from the main processing unit in a computing system. In hindsight, the use of the word "secure" in this term is a bit of a misnomer. Introductory lectures in computer security often rail against using the word "secure" in the absence of parameters such as "achieving what goal" and "against whom."

From the earliest days, secure coprocessors were envisioned as a tool to achieve certain properties of computation and storage, despite the actions of *local* adversaries—such as the operator of the computer system, and the computation running on the main processing unit. (Dave Safford and I used the term *root secure* for this property [SS01].) The key issue in secure coprocessors is

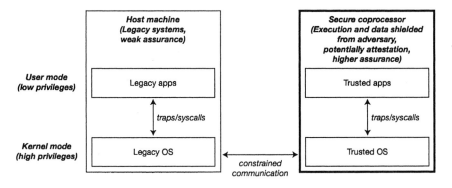

Figure 1.1. In the secure coprocessor model, a separate coprocessor provides increased protections against the adversary. Sensitive applications can be housed inside this protected coprocessor; other helper code executing inside the coprocessor may enhance overall system and application security through careful participation with execution on the main host.

not *security* per se, but is rather the establishment of a trust environment *distinct* from the main platform. Properly designed applications running on this computing system can then use this distinct environment to achieve security properties that cannot otherwise be easily obtained. Figure 1.1 sketches this approach.

Cryptographic Accelerators. Deployers of intensively cryptographic computation (such as e-commerce servers and banking systems) sometimes feel that general-purpose machines are unsuitable for cryptography. The modular mathematics central to many modern cryptosystems (such as RSA, DSA, and Diffie-Hellman) becomes significantly slower once the modulus size exceeds the machine's native word size; datapaths necessary for fast symmetric cryptography may not exist; special-purpose functionality, like a hardware source of random bits, may not be easily available; and the deployer may already have a better use for the machine's resources.

Reasons such as these gave rise to *cryptographic accelerators*: special-purpose hardware to off-load cryptographic operations from the main computing engines. Cryptographic accelerators range from single-chip coprocessors to more complex stand-alone modules. They began to house sensitive keys, to incorporate features such as physical security (to protect these keys) and programmability, (to permit the addition of site-specific computation). Consequently, cryptographic accelerators can begin to to look like trusted computing platforms.

Personal Tokens. The notion of a *personal token*—special hardware a user carries to enable authentication, cryptographic operations, or other services—

also overlaps with the notion of a trusted computing platform. Personal tokens require memory and typically host computation. Depending on the application, they also require some degree of physical security. For one example, physical security might help prevent a thief (or malicious user) from being able to learn enough from a token to create a useful forgery. Physical security might also help to prevent a malicious user from being able to amplify his or her privileges by modifying token state. Form factors can include smart cards, USB keyfobs, "Dallas buttons" (dime-sized packages from Dallas Semiconductor), and PCMCIA/PC cards.

However, because personal tokens typically are mass-produced, carried by users, and serve as a small part of a larger system, their design tradeoffs typically differ from higher-end trusted computing platforms. Mass production may require lower cost. Transport by users may require that the device withstand more extreme environmental stresses. Use by users may require displays and keypads, and may require explicit consideration of usability and HCISEC considerations. Use within a larger system may permit moving physical security to another part of the system; for example, most current credit cards have no protections on their sensitive data—the numbers and expiration date—but the credit card system is still somehow solvent.

Dongles.   Another variation of a trusted computing platform is the *dongle*—a term typically denoting a small device, attached to a general purpose machine, that a software vendor provides to ensure the user abides by licensing agreements. Typically, the idea here is to prevent copying the software. The main software runs on the general purpose machine (which presumably is at the mercy of the malicious user); this software then interacts with the dongle in such a way that (the vendor hopes) the software cannot run correctly without the dongle's response, but the user cannot reverse-engineer the dongle's action, even after observing the interaction.

Dongles typically require some degree of physical security, since easy duplication would enable easy piracy.

Trusted Platform Modules.   Current industry efforts center on a *trusted platform module (TPM)*: an independent chip, mounted on the motherboard, that participates and (hopefully) increases the security of computation within the machine. TPMs create new engineering challenges. They have the advantage of potentially securing the entire general purpose machine, thus overcoming the CPU and memory limits of smaller, special-purpose devices; they also let the trusted computing platform more easily accommodate legacy architectures and software. On the other hand, providing effective security for an entire system by physically protecting the TPM and leaving the CPU and memory exposed is a delicate matter; furthermore, the goal of adding a TPM to every commodity machine may require lower cost, and lower physical security.

Hardened CPUs. Some recent academic efforts seek instead to add physical security and some additional functionality to the CPU itself. Like the industrial TPM approach, this approach can potentially transform an entire general purpose machine into a trusted computing platform. By merging the armored engine with the main processing site, this approach may yield an easier design problem than the TPM approach; however, by requiring modifications to the CPU, this approach may also make it harder to accommodate legacy architectures.

Security Appliances. Above, we talked about types of devices one can add to a general-purpose machine to augment security-related processing. Other types of such specialized *security appliances* exist. For example, some commercial firms market hardened *network interface cards (NICs)* that provide transparent encryption and communication policy between otherwise unmodified machines in an enterprise. For another example, PC-based postal meters can also require hardened postal appliances at the server end—since a malicious meter vendor might otherwise have motive and ability to sell postage to his or her customers without reimbursing the postal service. Essentially, we might consider such appliances as a type of trusted computing platform pre-bundled with a particular application.

Crossing Boundaries. However, as with many partitions of things in the real world, the dividing line between these classes is not always clear. The IBM 4758 secure coprocessor platform drew on research into anti-piracy dongles, but IBM marketed it as a box that, with a free software application, the customer could turn into a cryptographic accelerator. (Nevertheless, many hardened postal appliances are just 4758s with special application software.) Some senior security researchers assert that secure coprocessing experiments on earlier generation IBM cryptographic accelerators predate the anti-piracy work. Some engineers have observed that current TPM chips are essentially smart card chips, repackaged. Other engineers assert that anything can be transformed into a PCMCIA token with enough investment; secure NICs already are.

1.3 Design and Applications

Many questions play into how to build and use a trusted computing platform.

Threat Model. Who are the adversaries? What access do they have to the computation? How much resources and time are they willing to expend? Are there easier ways to achieve their goal than compromising a platform? Will compromise of a few platforms enable systematic compromise of many more? Might the adversary be at the manufacturer site, or the software developer site, or along the shipping channel?

Deployment Model. A trusted computing platform needs to make its way from its manufacturer to the user site; the application software also needs to make its way from its developer to the trusted computing platform. The paths and players involved in deployment create design issues. Is the device a generic platform, or a specific appliance? Does the software developer also ship hardware? If software installation happens at the user site, how does a remote party determine that the executing software is trustworthy? Is the device intended to support multiple applications, perhaps mutually hostile?

More issues arise once the platform is actually configured and deployed. Should the platform support code maintenance? Can the platform be re-used for another application? Can an installation of an application be migrated, with state, to another trusted computing platform? Can physical protections be turned on and off—and if so, what does this mean for the threat model? Can we assume that deployed platforms will be audited?

Architecture. How do we balance all these issues, while building a platform that actually does computation?

Basic performance resources comprise one set of issues. How much power does the CPU have? Does the platform have cryptographic engines or network connections? More power makes life easier for the application developer; however, more power means more heat, potentially complicating the physical security design. User interfaces raise similar tradeoffs.

Memory raises additional questions. Besides the raw sizes, we also need to consider the division between types, such as between volatile and non-volatile, and between what's inside the physical protection barrier, and what lies outside (perhaps accessible to an adversary). Can a careful physical attack preserve the contents of non-volatile memory? What can an adversary achieve by observing or manipulating external memory?

Security design choices also interact with architecture choices. For example, if an on-the-motherboard secure chip is intended to cooperate with the rest of the machine to form a trusted platform, then the architecture needs to reflect the mechanics of that cooperation. If a general-purpose trusted platform is intended to persist as "secure" despite malicious applications, then we may require additional hardware protection beyond the traditional MMU. If we intend the platform to destroy all sensitive state upon tamper, then we need to be sure that all components with sensitive state can actually be zeroized quickly.

Applications. All these issues then play into the design and deployment of actual applications.

Is the trusted platform secure enough for the environment in which it must function? Is it economically feasible and sufficiently robust? Can we fit the application inside the platform, or must we partition it? Can we assume that a platform will not be compromised, or should we design the application with the idea that an individual compromise is unlikely but possible? How does the application perform? Is the codebase large enough to make updates and bug fixes likely—and if so, how does this mesh with the platform's code architecture? Will the application require the use of heterogeneous trusted computing platforms—and if so, how can it tell the difference? Finally, why should anyone believe the application—or the trusted computing platform underneath it—actually works as advertised?

1.4 Progression

In what follows, we will begin by laying out the big picture. Modern computing raises scenarios where parties need to trust properties of remote computation (Chapter 2); however, securing computation against an adversary with close contact is challenging (Chapter 3). Early experiments laid the groundwork (Chapter 4) for the principal commercial trusted computing efforts:

- High-end secure coprocessors—such as the IBM 4758—built on this foundation to address these trust problems (Chapter 5 through Chapter 9).

- The newer TCPA/TCG hardware extends this work, but enables a different approach (Chapter 10 through Chapter 11).

Looming industrial efforts—such as the not-yet-deployed NGSCB/Palladium and LaGrande architectures—as well as ongoing academic research explore different hardware and software directions (Chapter 12).

Chapter 2

MOTIVATING SCENARIOS

In this chapter, we try to set the stage for our exploration of trusted computing platforms. In Section 2.1, we consider the adversary, what abilities and access he or she has, and what defensive properties a trusted computing platform might provide. In Section 2.2, we examine some basic usage scenarios in which these properties of a TCP can help secure distributed computations. Section 2.3 presents some example real-world applications that instantiate these scenarios. Section 2.4 describes some basic ways a TCP can be positioned within a distributed application, and whose interests it can protect; Section 2.5 provides some real-world examples. Finally, although this book is not about ideology, the idealogical debate about the potential of industrial trusted computing efforts is part of the picture; Section 2.6 surveys these issues.

2.1   Properties

In its classic conception, a trusted computing platform such as a secure coprocessor is an armored box that does two things:

- It protects some designated data storage area against an adversary with certain types of direct physical access.

- It endows code executing on the platform with the ability to prove that it is running within an appropriate untampered environment.

What types of attacks the platform defends against, and exactly how code does this attestation, are issues for the platform architect.

In an informal mental model of a distributed computing application, we map computation and data to platforms distributed throughout physical space. Users (including potential adversaries) are also distributed throughout this space. Co-location of a user and a platform gives that user certain types of access to that platform: through "ordinary" usage methods as well as malicious attack methods (although the distinction between the two can sometimes reduce to how well the designer anticipated things). A user can also reach a platform over a network connection. However, in our mental model, direct co-location differs qualitatively. To illicitly read a stored secret over the network, a user must find some overlooked design or implementation flaw in the API. In contrast, when the user is in front of the machine, he or she could just remove the hard disk.

Not every user can reach every location. The physical organization of space can prevent certain types of access. For example, an enterprise might keep critical servers behind a locked door. Sysadmins would be the only users with "ordinary" access to this location, although cleaning staff might also have "ordinary" access unanticipated by the designers. Other users who wanted access to this location would have to take some type of action—such as picking locks or bribing the sysadmins—to circumvent the physical barriers.

The potential co-location of a user and a platform thus increases the potential actions a user can take with that platform, and thus increases the potential malicious actions a malicious user can take. The use of a trusted platform reduces the potential of these actions. It is tempting to compare a trusted platform to a virtual locked room: we move part of the computation away from the user and into a virtual safe place. However, we must be careful to make some distinctions. Some trusted computing platforms might be more secure than a machine in a locked room, since many locks are easily picked. (As Bennet Yee has observed, learning lockpicking was standard practice in the CMU Computer Science Ph.D. program.) On the other hand, some trusted computing platforms may be less secure than high-security areas at national labs. A more fundamental problem with the locked room metaphor is that, in the physical world, locked rooms exist before the computation starts, and are maintained by parties that exist before computation starts. For example, a bank will set up an e-commerce server in a locked room before users connect to it, and it is the bank that sets it up and takes care of it. The trusted computing platform's "locked room" can be more subtle (as we shall discuss).

2.2   Basic Usage

This discussion leaves us with the working definition: a TCP moves part of the computation space co-located with the user into a virtual locked room, not necessarily under any party's control. In more concrete terms, this tool has many potential uses, depending on what we put in this separate environment. At an initial glance, we can look on these as a simple 2x2 taxonomy: secrecy and/or authenticity, for data and/or code.

Since we initially introduced this locked room as a data storage area, the first thing we might think of doing is putting data there. This gives *secrecy of data*. If there is data we do not want the adversary to see, we can shelter it in the

TCP. Of course, for this protection to be meaningful, we also need to look at how the data got there, and who uses it: the implicit assumption here is that the code the TCP runs when it interacts with this secure storage is also trustworthy; adversarial attempts to alter it will also result in destruction of the data.

In Chapter 1, we discussed the difference between the terms "trustworthy" and "trustable". Just because the code in the TCP might be trustworthy, why should a relying party trust it? Given the above implicit assumption—tampering code destroys the protected data—we can address this problem by letting the code prove itself via use of a key sheltered in the protected area, thus giving us *authenticity of code*.

In perhaps the most straightforward approach, the TCP would itself generate an RSA key pair, save the private key in the protected memory, and release the public key to a party who could sign a believable certificate attesting to the fact that the sole entity who knows the corresponding private key is that TCP, in an untampered state. This approach is straightforward, in that it reduces the assumptions that the relying party needs to accept. If the TCP fails to be trustworthy or the cryptosystem breaks, then hope is lost. Otherwise, the relying party needs only needs to accept that the CA made a correct assertion.

Another public key approach involves having an external party generate the key pair and inject the private key, and perhaps escrow it as well. Symmetric key approaches can also work, although the logic can be more complex. For example, if the TCP uses a symmetric key as the basis for an HMAC to prove itself, the relying party must also know the symmetric key, which then requires reasoning about the set of parties who know the key, since this set is no longer a singleton.

Once we have set up the basis for untampered computation within the TCP to authenticate itself to an outside party—because, under our model, attack would have destroyed the keys—we can use this ability to let the computation attest to other things, such as data stored within the TCP. This gives us *authenticity of data*. We can transform a TCP's ability to hide data from the adversary into an ability to retain and transmit data whose values may be public—but whose authenticity is critical.

Above, we discussed secrecy of data. However, in some sense, code is data. If the hardware architecture permits, the TCP can execute code stored in the protected storage area, thus giving us *secrecy of code*. Carrying this out in practice can be fairly tricky; often, designers end up storing encrypted code in a non-protected area, and using keys in the protected area to decrypt and check integrity. (Chapter 6 will discuss this further.) An even simpler approach in this vein is to consider the main program public, but (in the spirit of Kerckhoff's law) isolate a few key parameters and shelter them in the protected storage.

However, looking at the potential taxonomy simply in terms of a 2x2 matrix overlooks the fact that a TCP does not just have to be passive receptacle that holds code and data, protected against certain types of adversarial attack. Rather, the TCP houses computation, and as a consequence of this protected environment and storage, we can consider the TCP as a computational entity, with state and potentially aware of real time. This entity adds a new column to our matrix: rather than just secrecy and authenticity, we can also consider *guarding*. Whether a local user can interact with the stored data depends on whether the computational guard lets him or her; whether a local user can invoke other computational methods depends on whether the guard says it is permissible.

2.3   Examples of Basic Usage

Secrecy of Data.   An axiom of most cryptographic protocols is that only the appropriate parties know any given private or secret key. Consequently, a natural use of TCPs is to protect cryptographic keys. A local user Bob would rather not have his key accessible by a rogue officemate; an e-commerce merchant Alice would rather not have her SSL private key accessible by an external hacker or a rogue insider.

Authenticity of Code.   Let's continue the SSL server example. Bob might point his browser to Alice's SSL server because he wants to use some service that Alice advertises. The fact that the server at the other end of the Internet tunnel proved knowledge of a private key does not mean that this server will actually provide that service. For example, Bob may wish to whisper his private health information so Alice's server can calculate what insurance premium to charge him; he would rather Alice just know the premium, rather than the health information. For another example, perhaps Alice instead is a healthcare provider offering an online collection of health information. Bob might wish to ask Alice for a record pertaining to some sensitive disease, and he would rather no one—not even Alice—know which topic he requested.

In both these cases, Bob wants to know more than just that the server on the end of the tunnel knows the private key—he also wants to know that the server application that wielded this data and provides this service actually abides by these privacy rules.

Authenticity of Data.   Suppose instead that Alice participates in a distributed computation in which she needs to store a critical value on her own machine. For example, we can think of an "e-wallet" where the value is the amount cash the wallet holds, or a game in which the value is the number of points that Alice has earned. We might even think more generally: perhaps this value is the audit log of activity (potentially from hackers) on Alice's machine.

In all these situations, the value itself might reasonably be released to Alice and to remote parties (under the appropriate circumstances). However, in these situations, parties exist who might have access to this value, and might have motivation to alter it. Alice may very well have motivation to increase her wallet and point score; an attacker who's compromised Alice's machine might very well want to suppress or alter the audit log. The remote party wants assurance that the reported value is accurate and current.

Secrecy of Code. Despite textbook admonitions against "security through obscurity," scenarios still arise in the real world where the internal details of a program are still considered proprietary. For example, credit card companies use various advanced data mining approaches to try to identify fraudulent account activity and predict which accounts will default, and regard the algorithm details as closely held secrets. Similarly, insurance companies may regard as proprietary the details of how they calculate premiums based on the information the applicant provided.

If Alice is such a party, then she would not want to farm her code out to Bob's site unless Bob could somehow assure her that the details of the code would not leak out. In this case, the TCP enables an application that otherwise might not be reasonable.

Guarded Data. In the e-wallet case above, Alice's TCP holds a register indicating how much money Alice's wallet holds. Consider how this value should change: it should only increase when the e-wallet of some Bob is transferring that amount to it; it should only decrease when Alice's e-wallet is transferring that amount to the e-wallet of some Bob. In both these situations, the exchange needs to be fully *transactional*: succeeding completely or failing completely, despite potential network and machine failures.

In this case, the relying party needs to do more than just trust that the value allegedly reported by Alice's e-wallet was in fact reported by Alice's e-wallet. Rather, the relying party also needs to be able to trust that this value (and the values in all the other e-wallets) has only changed in accordance with these transactional rules. By providing an authenticated shelter for code interacting with protected data, a TCP can address this problem.

For another case, consider an electronic object, such as a book or a movie, whose usage is governed by specific licensing rules. For example, the book may be viewed arbitrarily, but only on that one machine; the movie might have the additional restrictions of being viewed only N complete times, and only at ordinary speed. In both cases, a TCP could store the protected data (or the unique keys necessary to decrypt it), as well as house a program that uses its knowledge of state and time to govern the release of the protected object.

Of course, for this technology to be effective against moderately dedicated attackers, either the TCP needs to have an untappable I/O channel to release the material, or the material that is released during ordinary use must be somehow inappropriate for making a good pirated copy. (For one examples, we could use the TCP to insert watermarks and fingerprints into the displayed content.)

The notion of a protected database of sensitive information—where stakeholder policy dictates that accesses be authorized, specific, and rare—satisfies this latter condition. One example of such a database might be archives of network traffic, saved for later use in forensic investigation.

Guarded Code. As a natural extension to the above DRM example, we could change the book to a program—since the assumption that the adversary would not reverse-engineer the program solely from the I/O behavior observed during normal use is far more reasonable. In this case, the guard would prevent the program from operating—or migrating out of the TCP—unless these actions comply with the license restrictions. For the case in which the TCP is too limited in computational power to accommodate the program it is intended to protect, researchers have proposed *partitioned computation*: isolating a critical piece of the program that is hard to reverse-engineer, and protecting that piece inside the TCP.

A more trivial example would be a cryptographic accelerator: we do not want the TCP to just store the keys; we also want it to use the keys only when properly authorized, and only for the intended purpose. (As recent research shows, doing this effectively in practice, for current cryptographic hardware supporting current commodity PCs, is rather tricky.)

2.4   Position and Interests

Putting trusted computing protections in place for something that occurs only in one place involving one party does not achieve much. Arguably, TCPs only make sense in the context of a larger system, distributed in space and involving several parties. In the current Internet model, the initial way we think of such a system is as a local client interacting with a remote server. Typically, these terms connote several asymmetries: the client is a single user but the server is a large organization; the client is a small consumer but the server is a large content provider; the client handles rather little traffic, but the server handles much; the client has a small budget for equipment, but the server has a large one.

TCPs need to exist in a physical location, and to provide a virtual island there representing the interests of a party at another location. Initially, then, we can position a TCP in two settings:

- at the client, protecting the interests of the server,

- or at the server, protecting the interests of the clients.

However, like most things initial, this initial view misses some subtleties.

- Sometimes, a TCP at Alice's site can advance her own interests, much as a bank vault helps a bank. The TCP can help her protect her own computation against adversaries and insider attack. In e-commerce scenarios, this protection can even give her a competitive advantage.

- The client-server model may indeed describe much distributed computation. However, it does not describe all of it: for example, some systems consist instead of a community of peers.

- Naively, we think of a TCP as protecting some party's interests. However, the number of such parties does not necessarily have to be one.

- Naively, we also think of a TCP providing a protected space that extends the computational space controlled by some remote party. However, the number of parties who "control" the TCP's protected space does not necessarily have to be nonzero. E.g., if Alice is to reasonably gain a competitive advantage by putting some of here computation into a locked box, then the locked box must be subsequently under *no one's* control.

2.5   Examples of Positioning

Client-side.   The standard DRM examples sketched above constitute the classic scenario where the TCP lives at the client side and protects the interests of a remote server (in this case, the content provider). The operator of the local machine would benefit from subverting the protections, in order to be able to copy the material or watch the movie after the rental period has expired. Symmetrically, the remote content provider would (presumably) suffer from this action, due to lost revenue.

Server-side.   Above, we also sketched examples where the TCP lived at the server side:

- enforcing that access to archived sensitive data follows the policy agreed to before the archiving started; or

- providing a Web site where clients can request sensitive information, without the server learning what was requested.

These cases invert the classic DRM scenario. The TCP now lives at the server side and protects the client's interests by restricting what the server can do.

Protecting own interests.   This privacy-enhanced directory application also inverts the standard model, in that the TCP at the server side also arguably advances the server's interests as well: the increased assurance of privacy may draw more clients (and perhaps insulate the server operator against evidence discovery requests). Another example would be an e-commerce site that provides gaming services to its clients, and uses a TCP to give the clients assurance that the gaming operations are conducted fairly. By using the TCP to provide a space for fair play, the server operator advances her own interests: because more clients may patronize a site that has higher assurance of fairness.

We can also find examples of this scenario at the client. Consider the problem of an enterprise whose users have certified key pairs, but insist on using them from various public access machines, exposed to potential compromise. In one family of solutions, user private keys live in some protected place (such as at a remote server, perhaps encrypted). When Alice wishes to use her private key from a public machine, she initiates a protocol that either downloads the key, or (in one subfamily) has the machine generates a new key pair, which the remote server certifies.

In these settings, Alice is at risk: an adversary who has compromised this public machine can now access the private key that now lives there. However, suppose this machine used one of the newer TCP approaches that attempt to secure an entire desktop. We could then amend the key protocol to have the remote server verify the integrity of the client machine before transferring Alice's credential—which helps Alice. Thus, by using a TCP at the client to restrict the client's abilities, we advance the interests of the client.

Multiple parties. As we observed, the parties and protected interests involved can be more complex than just client and server. Let's return the health-insurance example. Both the client and the insurance provider wish to see that an accurate premium is calculated; the client further wishes to see that the private health information he provided remains private. Using a TCP at the insurance provider thus advances the interests of multiple parties: both the client and the server. We can take this one step further by adding an insurance broker who represents several providers. In this case, any particular provider might farm out her premium-calculation algorithm to the broker, but only if the broker can provide assurances that the details of the algorithm remain secret. So, a TCP at the broker now advances the privacy interests of both the consumer and the external provider, the accuracy interests of all three parties, and the competitive advantage of the broker.

For another example, consider the challenges involved in carrying out an online auction. Efficiency might argue for having each participant send in an encoding of his or her bidding strategy, and then having a trusted auctioneer play the strategies against each other and announce the winner. However, this approach raises some security issues. Will the auctioneer fairly play the strategies against each other? Will the auctioneer reveal private details of individual strategies? Will the auctioneer abide by any special rules advertised for the auction? Can any given third party verify that the announced results of an auction are legitimate?

We could address these issues by giving the auctioneer a TCP, to house the auction software, securely catch strategies, and sign receipts attesting to the input, output, and auction computation. The TCP here protects the interests of each of the participants against insider attack at the auction site and (depending on how the input strategies are packaged) against fraudulent participant claims about their strategies.

Community of peers. Consider the e-wallet example from earlier. If Bob can manage to increase the value of cash his e-wallet stores without going through the proper protocol, then he essentially can mint money—which decreases the value of everyone's money. In this case, the TCP at a client is protecting the interests of an entire community of peer clients.

Of course, the classic instantiation of such community-oriented systems is *peer-to-peer* computation: where individual clients also provide services to other clients, and (often) no centralized servers exist. Investigating the embedding of TCPs in P2P computation is an area of ongoing research. For example, in distributed storage applications that seek to hide the location and nature of stored items, using TCPs at the peers can provide an extra level of protection against adversaries. For another example, the *SEmi-trusted Mediator (SEM)* approach to PKI breaks user private keys into two pieces (using *mediated RSA*), and stores on piece at a trusted server, who (allegedly) only uses it under the right circumstances. We could gain scalability and fault tolerance by by replacing the server with a P2P network; using TCPs at the peers would give us some assurance that the key-half holders are following the appropriate rules.

No one in control. As we discussed above, in a naive conception, the TCP provides an island that extends the controlled computational space of some remote party. However, note that a large number of the above applications depend on the fact that, once the computational entity in the TCP is set up, no one has control over it, not even the parties whose interests are protected. For example, in the private information server, neither the server operator nor the remote client should be able to undermine the algorithm; in the auction case, no party should be able to change or spy on the auction computation; in the insurance broker case, the insurance provider can provide a premium calculation algorithm that spits out a number, but should not be able to replace that with on that prints out the applicant's answers.

How to build a TCP that allows for this sort of uncontrolled operation—while also allowing for code update and maintenance—provides many challenging questions for TCP architecture.

2.6   The Idealogical Debate

The technology of trusted computing tends to focus on secrecy ("the adversary cannot see inside this box") and control ("the adversary cannot change what this box is doing"). Many commercial application scenarios suggested for this technology tend to identify the end user as the adversary, and hint at perhaps stopping certain practices—such as freely exchanging downloaded music, or running a completely open-source platform—that many in our community hold dear.

Perhaps because of these reasons, the topic of trusted computing has engendered an idealogical debate. On the one side, respected researchers such as Ross Anderson [Anda] and activist groups such as the Electronic Frontier Foundation [Sch03b, Sch03a] articulate their view of why this technology is dangerous; researchers on the other side of the issue dispute these claims [Saf02b, Saf02a, for example].

Any treatment of TCPs cannot be complete without acknowledging this debate. In this book, we try to focus more on the history and evolution of the technology itself, while also occasionally trying show by example that TCP applications can actually be used to empower individuals against large wielders of power.

2.7   Further Reading

We'll consider many of these applications further in Chapter 4, Chapter 9, and Chapter 11.

Chapter 3

ATTACKS

A key component of trusted computing platforms is that they keep and use secrets, despite attempts by an adversary—perhaps with direct physical access—to extract them.

The broadness of the range of possible attack avenues complicates the task of addressing them. Contrary to popular folklore, one *can* sometimes prove a negative, if the space under consideration has sufficient structure. However, the space of "arbitrary attack on computing devices" lacks that structure. In the area of protocol design or even software construction, one can apply a range of formal techniques to model the device in question, to model the range of adversarial actions, and then to reason about the correctness properties the device is supposed to provide nonetheless. One can thus obtain at least some assurance that, within the abstraction of the model, the device may resist adversarial attacks. (Chapter 8 will consider these issues further.)

However, when we move from an abstract notion of computation to its instantiation as a real process in the physical world, things become harder. All the real-world nuances that the abstraction hid become significant. What *is* the boundary of this computational device, in the real world? What are the outputs that an adversary may observe, and the "inputs" an adversary may manipulate in order to act on the device?

These answers are hard to articulate, but designing an architecture to defend against such arbitrary attacks requires an attempt to articulate them. Some aspects follow directly from the considering the adversary.

- What type of access does the adversary have? Can he access the TCP while it is being shipped? Can he access it while it is dormant? Can he access it during live operation? If during live operation, how many of the operational parameters is the adversary free to choose? Are there any inherent or imposed limits on the number of adversarial operations?

- Is the adversary willing to try destructive analysis? How many units is he willing to destroy? How significant an advantage will the adversary gain by compromising few units? (E.g., does a given TCP contain a secret that would be useful in attacking the rest?)

- What tools and resources will the adversary bring to this problem? TCP designers commonly field such questions. "Could one break this with a $1 million budget?" "Could the NSA break this?" Such analysis can be useful, particularly when gauging how to allocate defense resources, in relation to the likelihood of the threat and the value of the target being protected. However, such analysis suffers uncertainty at the extremes. At the top end, those of us "outside the fence" of classified work can only speculate about the abilities such agencies have. At the bottom end, such analysis can suffer from underestimating the effectiveness of low-tech approaches. The defender must also keep in mind how easy it might be for an initially difficult attack to be transformed into something easily repeatable. (This phenomenon—innovative exploit turned into highly reproducible automated script—characterizes much of the known attack activity in the current Internet.)

Abraham et al [ADDS91] formalized the adversary space into three classes: clever outsiders, knowledgeable insiders, and funded organizations.

The architecture of the TCP also shapes its attack profile. Specific features and design choices can create their own adversarial opportunities. We consider some examples.

- If the TCP depends on an external device for resources like power or clocking, then these may become elements for the adversary to manipulate.

- Design goals and constraints may lead to a variety of types of memory in the TCP. For example, the informal model that Chapter 2 presented already introduced two variations: a protected area that the device either destroys or otherwise renders unavailable upon attack, and another area to retain non-sensitive data and code; an adversary cannot modify the latter without triggering the tamper response on the former. We might see further distinctions between volatile and non-volatile memory, and between dormant and run-time memory. Some architectures may even make use of storage outside the TCP.

Each type of memory raises its own issues. Do the alleged tamper protections actually work? Is the zeroized memory recoverable? Do run-time memory and operational registers receive the same protections as protected memory?

For external memory, are the contents susceptible to analysis or replay? (That is, can the adversary substitute a valid but outdated version of some unit?) Do the access patterns reveal any useful information?

■ Chapter 1 discussed a range of physical packaging for TCPs. These design choices affect the attack profile. For example, a TCP that leaves computation and memory exposed (such as TCPA/TCG-based platforms) permits more attacks than one that puts physical protection around the entire unit. A single-chip TCP requires different attack and defense techniques than a larger encapsulated module. A TCP intended to be carried in a user's pocket may need to withstand a broader range of environmental conditions, which may complicate defending against attacks that use extreme conditions.

Thus, considering the potential attacks that a TCP must resist requires surveying an essentially arbitrary space, refined by consideration of the adversary and the TCP architecture. Where is the perimeter? What can an adversary do? In this chapter, we'll approach this problem by surveying some of the attack avenues that, over the years, have proven fruitful; we will also extract some TCP design principles from this experience.

Section 3.1 considers physical attacks from outside the TCP. Section 3.2 considers software attacks, particularly if the TCP allows the adversary to insert code. Section 3.3 considers attacks possible via unforeseen I/O channels, created by the physical existence of the computation. Section 3.4 considers potential attacks due to undocumented functionality in TCP components. Section 3.5 considers the challenges involved in destroying sensitive memory. Section 3.6 considers attacks that emerge when the TCP is integrated into a larger system. Finally, Section 3.7 considers strategies for defense.

3.1 Physical Attack

In this section, we consider physical attacks from outside the TCP, that seek to actively penetrate or otherwise disrupt the internal device.

By physically attacking a TCP, the adversary hopes to subvert its security correctness properties somehow, usually by extracting some secret the TCP was not supposed to reveal. At first glance, the natural way to achieve this goal is the direct approach: somehow bypass the TCP's protections and read the data. As the following sections will elaborate, this direct approach can often prove rather successful.

However, a rather sophisticated family of indirect approaches has emerged, where the adversary instead tries to induce an error into the TCP operation via some physical failure; if the TCP continues to operate despite the error, it may end up revealing enough information for the adversary to reconstruct the secret. Researchers at Bellcore originally described this attack, in a theoretical context of inducing errors in cryptographic hardware that carried out the CRT implementation of RSA [BDL97]. This result generated a flurry of follow-on results, some of which became known as *differential fault analysis (DFA)*. These theoretical attacks eventually became practical and demonstrable [ABF+ 03, for example] and eventually earned the name *Bellcore attacks*, after the employer of the authors of the original paper.

In this section, we'll examine both the direct and indirect approaches. We'll consider penetrating devices with no armor (Section 3.1.1), single-chip devices (Section 3.1.2), and multi-chip devices (Section 3.1.3).

3.1.1   No Armor

Some TCPs make use of computing resources that are not protected by physical armor. For example, in the TCPA/TCG architecture, the trusted platform module chip (considered protected against the adversary) uses the rest of the machine for computation; for another example, sometimes microcontrollers will use external memory for additional storage, but encrypt the addresses and data/operands to hide operational details from the adversary.

Clearly, an adversary can tap and inject signals in exposed printed circuit boards, and often modify the circuits as well. (Indeed, this is how engineers debug hardware.) Besides such "logic analyzer and Xacto knife" attacks, the adversary can also make use of features that the hardware itself often provides. A memory card with *dual-ported RAM* can permit the adversary to change memory contents after the TCPA/TCG TPM checks it. Modern computer architectures also offer *direct memory access (DMA)* support that lets a peripheral work directly with memory, bypassing the CPU. Initially intended to improve operational efficiency, DMA has also been used to improve security by having a special peripheral check the main memory for corruption [PFMA04]. However, an adversary can use malicious DMA to bypass TPM checks.

Encrypting the busses does not necessarily help, due to the relatively small granularity of the space of instructions. Markus Kuhn [AK96] describes how, with some inexpensive lab equipment and patience, he was able to systematically break the bus encryption protections that the Dallas Semiconductor DS5002FP employed.

Even on allegedly armored devices, the adversary can sometimes do useful things by exploiting the unprotected nature of I/O channels. In the early days of smart-card-enabled telephones, adversaries could obtain free calls by using masking tape to cover the contact through which the phone debited the card. More recently, adversaries have installed "man-in-the-middle" keypads over the real keypads in automatic teller machines, in order to learn user PINs.

3.1.2   Single Chip Devices

Single-chip devices—particularly smart cards—have received much attention in the attacker community, perhaps due to the ubiquity of smart cards in low-end commerce applications (providing motivation), and the low cost (making experiment and destructive analysis feasible for a larger population). Anderson and Kuhn's work here [AK96, AK97] provides an enlightening (and entertaining) survey of the various techniques they found effective in practice.

Opening, probing, and reverse-engineering a chip has been an ongoing cat-and-mouse game between the adversary and vendors. Once the adversary opens a device, he can probe EEPROM state and logic design. Selective alterations can also prove useful: the adversary can disable future changes to EEPROM values by destroying the capacitors that provide the EEPROM write voltage; the adversary can also re-join fused links and put the device back into factory state; the adversary can use UV light to selectively change EEPROM cells; the adversary can even make small alterations to chip logic.

The adversary can also manipulate the environment. Deviously abnormal supply voltages can clear critical critical EEPROM bits or force random number generators to generate mostly ones. The need for smart cards to tolerate varying clock rates can enable the adversary to make analysis easier by single-stepping the card. Appropriately timed and crafted transients in clock rate or supply voltage can induce a predictable disruption the CPU's execution of a selected instruction; the adversary can use this to disrupt control flow—e.g., turning a secure multi-round cipher into an easily breakable one-round cipher—or to carry out DFA attacks. Bar-El et al [BECN+ 04] discuss additional environmental avenues, such as applying low temperatures that affect the correctness of only some of the operations, and using lasers, X-rays and ion beams to induce errors. (Bar-El et al also include photographs of of their fault injection equipment.) Skorobogatov and Anderson recently described how to use camera flash devices to do fault-injection attacks on single-chip devices [SA03].

3.1.3   Multi-chip Devices

Multi-chip modules provide a different set of attack scenarios, as a larger device can be self-powered and use stronger materials.

Some early modules used defense techniques such as environmental sensors that determined when the outer armor was opened up: for example, a maintenance door might have a microswitch, or the internal device might have a light sensor. Fairly straightforward techniques may defeat these mechanisms: for example, the adversary may drill a hole in the door, and (through this hole) apply glue to keep the microswitch closed.

Early secure coprocessor designs suggested wrapping very thin wire around the module, and then dunking the result in epoxy-like resin. The hypothesis:

to penetrate the device, the adversary must break the wire—which an on-board circuit can notice. However, it turned out that patient adversaries with nimble hands could indeed unwind the wire—a strategy that my colleague Steve Weingart terms the "brain surgery" attack [Wei00]. Other sometimes effective techniques include machining with water, laser, and sand-blasting. Weingart also speculates on the use of shaped-charge explosives to create a high-speed plasma lance that can penetrate a device before defense circuitry can notice; unfortunately, when we worked together at IBM, he had not been able to find the appropriate explosives on the open market to try this.

Govindavajhala and Appel at Princeton recently published a novel way to use physically-induced errors to subvert language-based security mechanisms in general-purpose computers [GA03]. One way for programmers to produce system software that avoids common software security flaws (see Section 3.2 below) is to use a *type-safe* language, such as Java, that does not permit a program to store data in some memory object unless their types match. However, these languages check the type matching at build-time, rather than run-time. The Princeton researchers wrote a valid Java program that fills memory with a deviously constructed data structure such that, for a majority of the bits, if hardware error causes one bit to flip, then the program now has an integer and an integer-pointer living at the same address. The adversarial program can exploit this state to arbitrarily re-write memory, within this type-safe program that is supposed to disallow that.

The Princeton researchers then used *heat*—a light bulb next to the machine— to induce such hardware errors.

3.2   Software Attacks

Computing platforms that provide services through computational interfaces— e.g., function calls, network protocols, etc.—have a long history of permeability. TCPs are no exception. In the focus to think about physical attacks and defenses, a designer can easily overlook these software avenues. These susceptibilities historically fall into several classes; we'll quickly review them here.

It is important to note that these vulnerabilities appear to be fundamentally endemic to sufficiently large software systems. (How to build systems that avoid these risks remains an area of active research.) Without further countermeasures, a TCP that uses such software modules or depends on a commodity operating system for protection almost certainly suffers from such weaknesses. Also, even if the vulnerability occurs between two internal TCP modules–rather than on the outside, between the TCP and the adversary—it still may be possible for the adversary to exploit it, if the adversary can figure out how to trick the calling module to trigger this behavior.

3.2.1 Buffer Overflow

Computers traditionally store data in a buffer: a consecutive sequence of memory locations, identified by the address of the "zeroth" location. If a computing device receives input data from a user, it usually copies this data into such a buffer: byte for byte, starting with this zeroth location. If the data provided is longer than the buffer—and the system does not notice—then the system may blindly copy the remainder of the data into memory locations beyond the end of the buffer.

This *buffer overflow* overwrites whatever data the system had stored in these locations. If this location contained data critical to the correct operation of the system, then an adversarial user now has a chance to subvert correct operation by providing maliciously crafted, overly long input.

One primary way adversaries exploit buffer overflow is if the buffer lives on the execution stack. In this case, a sufficiently long input can rewrite the return address in the stack frame, and thus trick the device into "returning" to code of the adversary's own choosing—perhaps even code that the adversary himself *injected*, via the input. This style of attack is often called *stack smashing*. The victim device executes this code with whatever privileges the device had when it accepted the input; if a user process exploits a buffer overflow in kernel-level OS code, the user can execute instructions at kernel privilege.

An adversary can also exploit buffer overflow without rewriting addresses, if other particularly interesting state variables live in the overwritten area.

Buffer overflow attacks can exhibit considerable subtlety. For example, *return-to-libc* attacks can permit an adversary to (essentially) run arbitrary code, even if the device refuses to execute code that lives on the stack.

In some sense, buffer overflow is a solved problem, in that many software and hardware techniques have been proposed and prototyped to defend against it. Nonetheless, the problem remains ubiquitous in the field.

3.2.2 Unexpected Input

When a designer programs a system to accept and process input, he or she does this in the context of a particular operational sequence the system is carrying out, and in the context of a particular type of response the system should be expecting at this point. However, the input provided might essentially be any arbitrary byte string, not necessarily of the expected type (e.g., a pair of positive integers), nor of a legal element of that type (e.g., a valid account number, and a value of cash, not more than the account's current balance).

Designers tend to focus efforts on building the system so it works correctly if the user provides correct input for the context: legal data, of the legal types. This focus can lead the designer to overlook what may happen if the user provides input that does not meet these expectations; if such input can take the system into usefully corrupt states, an adversarial user can use this avenue to attack.

The standard defense against such attacks is *argument validation*: having each receiver of user input check that the input is valid for this context before acting on it. Nonetheless, it can be difficult to do this correctly for all cases, given the rush of commercial software production, the difficulty of formally specifying "correct" input, and the current lack of ubiquitous language support for validation.

Example: IBM 4758 Code-load Requests. In the IBM 4758 TCP, we permitted an external user to request that the device load a user-provided block of executable code into a certain FLASH sector. Should this request be properly authorized, the device would carry it out.

While it is pending, this code-load request lives in operational DRAM within the TCP; this DRAM also houses some critical device secrets. In our protection model, the TCP destroys this data upon tamper, but leaves the FLASH contents available to an adversary. To avoid buffer overflow risks, we checked that the command the user provided fit within the DRAM buffer we had allocated, before the TCP brought the command in. In order to accommodate a complex command structure with several varying-length items, we structured the command with an index area, with a pair of integers—for "offset" and "length," respectively—for each field. This pair indicated where the field could be found: "offset" indicated how many bytes away it was from the command area.

However, a devious adversary (who was authorized to provide a low-privileged code load for that TCP) could have constructed a command where the "code" offset sent the TCP out of the command buffer, all the way to the part of DRAM where the critical platform secrets lived. The consequence would have been that the TCP would dutifully copy the secrets into FLASH, where they would be available after physical penetration.

Even though the input was structured correctly and fit within the buffer, it was illegal: the user could reach beyond his or her confines, with input that did not satisfy validity constraints.

(We avoided this problem, by carefully testing that all such offsets lived within the input buffer, which had been cleared before bringing the input in.)

3.2.3   Interpretation Mismatches

Another common software vulnerability occurs when a system uses two different ways of mapping some sequence of bytes to its semantic meaning. Suppose module M_1 interprets user-provided data D as $F_1(D)$, but module M_2 uses some different F_2. If the designer does not take this difference into account, the adversary might be able to provide data D where $F_1(D) = F_2(D)$, and the modules would the carry out inconsistent actions, taking the system into an insecure state.

Such flaws commonly occur when systems process user-provided character strings, but different system modules use different interpretations of what character sequence denotes the end of the string.

For example, in order to ensure that external clients could only see html files, the eXtropia WebStore used a CGI script to check that a requested URL terminated in the string "html" before passing the request to the underlying OS. The script interpreted the NULL byte (0x00) as a valid character, but the OS interpreted it as the end of the string; consequently, an adversarial user could obtain any file, simply by appending a NULL byte and "html" to its name.

Because such interpretation mismatches commonly occur because of improper handling of such "escape" characters, this family is often called *escape sequence* flaws.

3.2.4 Time-of-check vs Time-of-use

Security systems already have an implicit notion of correctness predicates for actions. The system should not carry out a requested action unless the system's security policy permits that action, by that user, in that particular system state. The software processing these requests typically check that these conditions are satisfied. As the above discussions showed, security software may have many other correctness conditions that also need to be checked, if we're going to keep the system from entering an incorrect (and insecure) state.

However, the checks and the actions resulting from these checks do not typically take place in the same instant. Duration can exist between the check and the action. Duration can also exist in the action itself; sometimes, the check can be sufficiently complex that duration exists throughout the check. Such duration introduces the possibility of error and, perhaps, attack: the predicate may have been true when the system checked it, but may have ceased to be true when the system acted on the results. The check itself may have been based on assumptions and bindings that ceased to hold during the check process. The term *time-of-check/time-of-use (TOCTOU)* denotes this family of flaws.

Example: *Hamlet*. For a literary example, consider the fate of Rosencrantz and Guildenstern. To kill Hamlet, the King authorized the death of the bearer of a letter given to Hamlet; however, between the authorization and the action, the binding between "bearer of letter" and "Hamlet" ceased to hold.

Example: CP/Q++ Crypto Services. In the IBM 4758 TCP configured with the CP/Q++ operating system, the OS provides a suite of services to user-level processes. These services include requests for cryptographic services and for retrieval of sensitive data. The security modules within CP/Q++ used the message-passing paradigm (native to the CP/Q kernel) to implement these services, which typically took duration to carry out. The CP/Q kernel used a process identifier to indicate message recipients. The security modules checked that the caller was authorized, but embodied the results of this check as the process identifier, to which the response should be directed. However, between the check and the completion of the request, the caller may have terminated, and the kernel may have re-used its identifier for a new process—which receives a result message potentially containing data is not authorized to see.

We addressed the problem by putting all user-level processes into the same security domain.

3.2.5   Atomicity

A more direct consequence of duration can be a lack of *atomicity*. Suppose a TCP action takes non-trivial duration; rather than transitioning instantly from state A to state B , the TCP proceeds through a sequence of intermediate states $I_1, ...I_k$. If the adversary has the ability to interrupt the action (e.g., by removing power from a TCP that requires its outside environment to provide power), then the adversary may be able to cause the TCP to stop in one of these intermediate states I_i. If the designer did not anticipate this attack, resuming operation in state I_i may put the TCP into an insecure state.

For example, consider the action of erasing and rewriting a FLASH sector (see Section 3.4.2). Between the start and finish of this action, the contents of the sector are indeterminate; if the adversary can arrange the interruption to be after the sector has been erased but before the rewrite starts, the contents will be a known (erased) state. If the TCP stored critical data here and had no other non-volatile memory, then adversarial interruption could leave this data in an illegal state.

More traditional systems—such as databases and distributed transaction systems—have well-developed theories of atomicity and techniques to achieve it: to ensure that as far anyone can observe, actions either fail completely (the system remains in state A) or succeed completely (the system transitions to B). However, these techniques do not always easily adopt to some TCP architectures, where some actions may not be easily rolled back, and stable storage itself does not necessarily have atomicity.

Example: Gemplus Audit. Bar-El et al [BECN[+]04] discuss an EEPROM atomicity attack that Gemplus discovered in one of their DES smart cards in 1994. The adversary would like to learn the card's secret key k_0. First, the adversary asks the card to encrypt a value, so the adversary has a plaintext-ciphertext pair p_0, c_0. At that time, brute-force search of the full 56-bit DES keyspace was considered infeasible.

However, the adversary could initiate an erase of the EEPROM that contained the key but remove power after the EEPROM erased just the first block. Since the EEPROM had a block granularity of 32 bits, this interruption left the card with half the key bits reset to zero. The adversary then requests a second encryption, getting a p_1, c_1 pair: encrypted with a key k_1 consisting of 28 bits from k_0 and 28 zeros, in known places. The adversary does a 2^{28} brute-force search to learn k_1, then uses these bits to do another 2^{28} brute-force search to learn the rest k_0. Lack of atomicity in the EEPROM erase operation reduced an infeasible 2^{56} search to a feasible 2^{29}.

3.2.6   Design Flaws

These standard families of flaws can be highly effective in subverting security. This effectiveness tempts an analyst to look for such exploits when evaluating a system's security. However, as a consequence of focusing on how the system might incorrectly implement its interfaces, the analyst can overlook flaws that result from simply using the system's advertised functionality in unexpected ways. Software systems—and TCPs are no exception—can suffer from design flaws.

In recent years, researchers such as Mike Bond and Ross Anderson at Cambridge [BA01] have focused attention on the API level of these devices. Besides physical properties, instantiation of abstract ideas in the real world also can lead to feature creep. As Mike Bond paraphrases Needham, clean abstract designs tend to become "Swiss Army knives." In particular, cryptographic accelerators have found major commercial application in banking networks: for ATM and credit card processing, devices need to transmit, encode, and verify PINs. However, the accumulation of usage scenarios leads to a transaction set complexity that permits many clever ways to piece together program calls that disclose sensitive PINs and keys. Jolyon Clulow [Clu03] (formerly of Prism, but now also at Cambridge) has discovered many amusing attacks possible from exploiting error behavior resulting from devious modifications of legitimate transaction requests.

Besides feature creep, one conjectures that another source of these vulnerabilities is the inevitable complexity of software evolution paths in commercial projects. Projects split into branches, for different customer bases or contexts; these branches evolve separately; these branches then merge as a new variation tries to accommodate these various legacy uses. These mergers of different APIs can cause trouble: each alone might keep the system in a secure state, but their union can reach dangerous states, because one's specializations were never considered in the context of the other's.

3.3   Side-channel Analysis

Section 3.2 considered attack avenues that TCPs inherit by being, in part, software. These computing platforms are also physical. The physical action of computation can often result in physical effects an adversary can observe; these observations can sometimes betray sensitive internal data the TCP architecture was supposed to protect.

This style attack of is often called *side-channel analysis*, since the TCP leaks information via channels other than its main intended interfaces. In this section, we discuss these attacks.

- Section 3.3.1 discusses extracting information from the time computation takes.

- Section 3.3.2 discusses extracting information from power consumption; and

- Section 3.3.3 discusses some other avenues that have proved fruitful.

According to rumors, the U.S. Government's classified *TEMPEST* project has developed an extensive suite of knowledge and defenses in this vein. However, almost by definition, details of classified projects can be difficult to confirm, although some details have begun to be released (see http://cryptome.org/nsa-tempest.htm, or Chapter 10 in [RG91]).

3.3.1   Timing Attacks

Computation takes time: the CPU needs to fetch instructions and data; on a finer-grained level, gates must switch and wires must carry signals. The exact combination and sequence of actions and signals depends on the operational data, and the duration depends on this combination. But if the actions depends on secret data, the duration can betray this information.

Example: TENEX.   For a classic example of this approach, let's look back to password checking in the Tenex operating system, an early 1970s timesharing system for the PDP-10. Naively, the number of attempts necessary for an adversary to guess a secret password is exponential to the password's length. (E.g., for an 8-character password chosen over an alphabet of size N , it may take as many as N^8 guesses.)

However, Tenex made it much easier for the adversary because it checked a guess one character at time, and stopped at the first mismatch. By lining up a guess across the boundary between a resident page and a nonresident page and observing whether a page fault occurred when the system checked the guess, an adversary could verify whether a specific prefix of a guess was correct. For example, the adversary could begin by making a random guess as to what the password might be—but line this guess up along a page boundary so only the first character is in a resident page. If the system page-faults before rejecting the password, then the adversary knows that the first character in the guess was correct. Thus, it takes at most N guesses to get the first character, then N for the second, and so on, giving $8N$ in all—which is much better (for the adversary) than N^8.

The fact that the secret password comparison occurred on a real machine led to an observable property that turned an intractable exponential search into a feasible linear one.

Example: RSA. If we fast forward to 1995, the same basic problem—an observable artifact lets an adversary verify the prefix of a guess—emerged for more abstract cryptographic devices.

For example, consider a TCP that carries out modular exponentiation with a secret exponent, as part of the the RSA cryptosystem. The time that the modular operation $x^d \bmod N$ takes depends on the operand x, the modulus public N, and the secret exponent d; the operation and its standard implementation are well-understood by the community, including the adversary. Suppose the adversary can measure the time the TCP takes for some known x and N, and guesses a value g that might be the n-bit secret exponent d. The adversary can then calculate a prediction for the time the TCP should take for this operation, were the guess g the correct exponent. If only the k most-significant bits of the guess were correct, then the adversary's prediction model would be correct for these first k bits, but wrong for the remainder. Furthermore, the cryptographic operations involved essentially randomize the data the adversary's model will use for these remaining bits—so the timing components these bits introduce look more or less like statistically random variables.

For a given guess g, over enough samples x, the difference between predicted and real times would then form a statistical distribution with variance proportional to $n - k$. As a consequence, the adversary could confirm the correctness of a guessed k-bit prefix of the secret by running enough samples, measuring these differences, and calculating the variance. With enough samples, this artifact of the physical implementation of RSA (and other cryptosystems) turns an infeasible exponential search into a feasible linear one. Instantiating the cryptography in the real world leads to threats that do not always show up on a programmer's white board.

One defense approach is to make the operational parameters independent of the input data. The feasibility of this approach depends on the operation. For example, in RSA, one can use random data to conduct a *blinding* transformation on the parameters before the operation, and then a reverse unblinding transformation afterwards. However, carrying out this approach on a TCP that does not have a good source of randomness—or a good way to obtain a seed and store a context of pseudorandomness—can be tricky.

An easier defense approach—and one that newer-generation modular exponentiation and RSA engines started to incorporate—is to design the hardware to take constant time for each operation, no matter what the data was. When Paul Kocher first published his timing attacks in 1995, at least one old-timer claimed that a few older commercial accelerators also took constant time, indicating that some in the commercial world must have already known about the attack.

Reproducing the RSA timing attack has made an excellent homework project here at Dartmouth.

Example: Apache SSL Web Servers. In the classic instantiation of cryptographic timing attacks, the adversary has direct access to a TCP carrying out RSA, and the TCP uses some variation of the standard multiply-and-square implementation of the modular exponentiation step. However, in 2003, researchers from Stanford demonstrated that the adversary can even carry out such attacks from the other side of the network, against platforms—such as Apache with mod_SSL—that use much more sophisticated implementations [BB03]. All it requires is sufficiently many samples to overcome the noise the distance introduces, and sufficient cleverness in teasing apart the implementation details.

Example: Web Caches. Researchers at Princeton discovered another amusing example of timing attacks. When Alice visits Bob's Web site, her browser directs her machine to load and render the various elements that Bob's page specifies—except the browser first looks in Alice's local caches, since retrieving an object from there is quicker than retrieving it from a remote server. If an adversary can get Alice to visit a Web page that loads an element specific to Bob's page—and also times how long it takes for Alice's browser to do this— then the adversary can learn whether Alice has visited Bob's site recently.

Reproducing this cache-timing attack has also made an excellent homework[1] project.

Example: University Passwords. Here at Dartmouth, a university-wide name-and-password system handles authentication for mail, and for other services such as registering for classes, checking grades, and (for faculty and staff) accessing student records. To avoid sending passwords in plaintext, the authentication system adopted the "Random Number Exchange" technique from AppleShare: the client concatenates the ASCII encoding of the characters in the user's password to form a DES key, and then uses this key to encrypt an

---

[1] My forthcoming *Educause Quarterly* essay expands on carrying out fun security attacks as part of homework assignments [Smi04].

eight-byte random challenge from the server. (If the password is shorter than eight characters, the client pads it with zero-filled bytes.)

While exploring password hacking, we noticed that, on the wireless network, we could observe the campus authentication server send a random challenge to a nearby user's client, and then observe the client respond. We further noticed that the duration of the DES operation was proportional to the password length (perhaps because of the zero-filled bytes—this bears further investigation). Thus, by timing this interval, we could determine which users had passwords short enough for a quick brute-force search; by recording the challenge-response pair for such users, we could carry out that search.

However, since the only users on campus with short passwords were high-level faculty and administrators who chose their passwords before the length limit was in place (and who have resisted pleas to choose new passwords), we decided against pursuing this experiment further.

3.3.2 Power Attacks

In addition to the time-of-operation operation approach, physical devices have other observable physical characteristics that depend on hidden secrets. One natural characteristic is power. When complementary metal-oxide semiconductors switch, they consume power; an adversary could measure this consumption and attempt to deduce things about the operation.

SPA. With *simple power analysis (SPA)*, an adversary tries to draw conclusions from simple power traces. With sufficient knowledge of what the device does, the adversary can identify a particular point in the sequence where what the device does (and the power it consumes) will differ depending on what the hidden data is. Even with insufficient knowledge of the device architecture (or the operational parameters), the adversary can often still achieve results with a bit of experimentation. On the other hand, once the designer recognizes this attack avenue, defense can be fairly straightforward: as with timing defenses, the designer de-correlates the operational parameters from the secrets.

Example: a Lost Bet. SPA can be quite effective. Two co-workers of mine made a bet on how many power traces would be necessary to extract a DES key from a certain commercial smart card. The winner managed to do it with a single power trace: an initial parity check led to a mix of spikes, some short, some tall, one for each bit in the key.

DPA. More advanced *differential power analysis (DPA)* looks at subtle statistical correlations between the secret bits and power consumption.

In its classic instantiation, the adversary collects a large set $\{T_i, C_i\}$ of trace-ciphertext pairs. The adversary also picks a selection function D that takes a ciphertext and a guess of part of the key and outputs one bit. The idea is that if the guess is right, this bit reflects something that actually shows up in the computation, but if the guess is wrong, then D will be random across the ciphertexts.

The adversary then makes a guess K_g, and uses this guess and the selection function to partition the set of traces into two sets: the T_i for which D (C_i, K_g) = 0 and the T_i for which D (C_i, K_g) = 1. He averages the traces in each set, and then looks at the difference between these average traces. If K_g was wrong, these two sets are uncorrelated, and the differential trace becomes flat as the sample size increases. However, if K_g was right, the differential approaches the correlation of D and power consumption, which will be spiky.

Defenses against differential power analysis are difficult, since they essentially only reduce the signal the adversary is reading, rather than eliminate it.

3.3.3   Other Avenues

When Kocher first announced his timing attacks, a senior colleague of mine harrumphed and suggested many other physical avenues—including power, electromagnetic radiation, and heat—by which a TCP could leak information. Most of the avenues he suggested—and several he did not even imagine—have turned out to be feasible. (The sole exception: the conventional wisdom to date is that heat analysis is not effective [QS02].)

Electromagnetic.   As electrical devices, the components of a computer generate electromagnetic radiation as part of their operation. An adversary that can observe these emanations and can understand their causal relationship to the underlying computation and data may be able to infer a surprising amount of information about this computation and data. This ability can be devastating, should the computer be a TCP intended to keep this information from the adversary.

Defense against side-channel analysis via electromagnetic radiation is reputed to be the center of the classified world's TEMPEST project; one source dates this from the 1950s [RG91]. Knowledge of this attack avenue grew slowly in the public domain. According to one account [Hig86], a 1967 conference presentation was the first public disclosure; a 1983 document (*Läkande Datorer*) published in Swedish also discussed it. The breakthrough public publication was van Eck's 1985 paper [van85], discussing (and demonstrating) the ease of detecting the electromagnetic radiation from a CRT monitor and reconstructing what the monitor displayed—even from a distance of several hundred meters.

Kuhn and Anderson later followed up this work with *soft tempest* [KA98]. If the adversary can also send software to the target machine, he can craft software that deliberately enhances the signal strength in this side channel— and then uses this channel to transmit interesting data harvested from the victim.

Alternatively, a TCP designer can incorporate specially crafted *Tempest fonts* in what the CRT displays, to greatly reduce what a remote adversary can learn. In recent years, Kuhn has demonstrated [Kuhar] that modern flat-panel displays can also be vulnerable to this type of attack, and suggests using randomness in the bitmap each time a character is displayed.

In the above EM schemes, the adversary exploited the relatively strong signals generated by the CRT. Internal computational logic also generates signals, albeit more subtle ones. In follow-on to power analysis work, researchers [QS01, AARR, for example] demonstrated that similar techniques based on EM can also extract secrets from commercial devices such as chip cards and cryptographic accelerators. More recently, researchers have also examined the potential for using multiple side-channels simultaneously [ARR03].

Visible Light. Computing devices use light to communicate information to human users: think of a CRT monitor, or even status LEDs on modems and network ports. These devices directly communicate information via line of sight; however, their signals also indirectly affect things like diffuse reflection off of background objects. Markus Kuhn demonstrated [Kuh02]—via both sophisticated analysis as well as direct experiment—that the average luminosity of a CRT's diffuse reflection off of a wall can sufficient to reconstruct the signal displayed on the CRT (so shielding the CRT to protect against leaking information via electromagnetic radiation may not be sufficient). Kuhn also speculates that the same techniques can apply to LED signals. Even without line of sight, the adversary may be able to read the signals that a TCP's optical output channels emit.

Acoustic. Very recently, researchers [ST04] have demonstrated a preliminary proof-of-concept that a correlation exists between the *sound* of a processor and its computation.

3.4 Undocumented Functionality

TCPs, like other computing devices, are typically assembled from commodity components, such as ICs containing fairly complex semiconductor circuits. Like other complex systems, these devices offer a rich array of services, which the rest of the TCP may access via the component's external interface. This interface typically has a well-defined specification describing what services the component offers, and exactly how the interface should be manipulated to invoke these services. Designers turn this specification into a mental model that guides how they integrate the component into the broader TCP design.

However, like any other complex system, this model may diverge from reality. The component may have behavior that explicitly diverges from the documented specification; the component may also offer additional functionality that can be accessed via interface methods that do not appear in the specification. For example, a microprocessor executes its instructions by decoding the opcode, and then carrying out the specific operations required by the instruction this opcode represents. In *random logic* implementations, the microprocessor carries out this decoding by a web of logic gates—but, as a consequence, one might find that opcodes that do not represent valid instructions still cause the microprocessor to do something interesting and perhaps useful. In the early days of microprocessors, we hobbyists excitedly exchanged folklore about things such as undocumented instructions for certain vendors' 6502 chips. ("You can load two registers simultaneously!")

Under some circumstances, the adversary may be able to turn such divergences into attack avenues. This can occur when the divergence is relevant to the TCP's enforcement of its security properties, and when the adversary can cause (or wait for) the TCP to take this internal component into that behavior space.

Let's consider some real examples.

3.4.1   Example: Microcontroller Memory

A *microcontroller* is a small, single-chip computer—processor, RAM, code space, EEPROM non-volatile storage. In the IBM 4758, we used a particular vendor's microcontroller as part of our larger multi-chip physically encapsulated TCP. The microcontroller behavior the TCP accessed included asking the microcontroller to write critical values to its internal EEPROM. Among the behaviors an external TCP user (including the adversary) accessed included resetting the entire platform by interrupting power, and resetting the entire platform simply by leaving the device powered up but triggering the reset hardware line.

One might wonder what would happen should one of these TCP reset triggers occur while the internal microcontroller is carrying out a write to its EEPROM. For example, suppose the microcontroller was trying to store value 0x23 to address 0x1F, which currently stored value 0xA6. The natural mental model (and the one the documentation suggested) would imply that the EEPROM would be in one of two states:

- The write succeeds: address 0x1F now stores value 0x23.

- The write fails: address 0x1F still contains value 0xA6.

Paranoid designers might even suspect a third possible state:

- The write partially succeeds: address 0x1F now contains some unknown value (because the EEPROM was being changed while reset happened).

One naturally assumes that, in any of these cases, the remainder of EEPROM would be unchanged.

However, while doing some exploratory testing, one of our engineers tried a power-preserving platform reset during a microcontroller EEPROM write, and discovered that what actually happened was something unexpected:

- The write succeeds, but only after the destination address is reset to 0x00. Consequently, the EEPROM now has overwritten address 0x00 with 0x23.

The engineer contacted the vendor, who knew about this flaw but had neglected to document it. Until this point, we had been storing some critical flags in EEPROM location 0x00. Causing that location to be rewritten with an arbitrary value might have greatly benefited an adversary; for example, it might have convinced the TCP to turn back to "factory mode" with greatly reduced protections.

3.4.2 Example: FLASH Memory

FLASH is a type of non-volatile semiconductor memory, typically packaged in a single chip. In ordinary read mode, it appears to the rest of the system as typical ROM: the system uses the chip's address lines to specify a memory location, and then uses the chip's data lines to read the value stored there.

Changing the stored values is a more complicated operation: the system must first *erase* the memory (e.g., resetting the bits to zero), then put the memory into *write* mode, then write the values, then put it back into *read* mode. Each of these operations takes non-trivial duration; some programmers try to streamline things by using careful data structures that reduce the need for erase operations. FLASH chips typically partition their memory into large *sectors* that can be erased separately. To allow a single chip to provide some ROM storage as well as rewritable storage, chips frequently provide a way to make sectors read-only by board-level signals, such as grounding a particular pin on the IC.

Putting a FLASH chip into erase or write mode thus requires telling the chip to change modes, and telling it which sector to operate on. Typically, the interface permits this by having the system send a special sequence of "write" operations to a magic series of addresses. Consequently, the chip needs some decoding logic on its input lines, to recognize and process the "opcodes" that these sequences represent. The existence of this logic creates the potential for the existence, by accident or by design, for opcodes that carry out other, undocumented operations.

For example, a colleague[2] reported discovering FLASH chips with an undocumented operation sequence that would enable the system to erase and rewrite a sector that was supposed to have been hardwired as ROM. He conjectured that the vendor provided this feature deliberately, as a service to customers who discovered, after installing the devices, they needed to change the ROM code.

---

[2]It gives me great pleasure to note that this colleague's last name was "Hack." I kid you not!

However, a TCP designer may very well use ROM protection on FLASH sectors as part of the security protections—for example, if the TCP permits external users to provide code that runs in some protected way, the designer may trust the ROM protections to preserve boot-block code from adversarial modification. If the adversarial code has write access to other sectors on that FLASH chip, then (without further countermeasures) the adversary could use such an undocumented feature to defeat the TCP security.

3.4.3   Example: CPU Privileges

In contemporary computer architecture, CPUs operate in one of a number of privilege modes. Internal hardware restricts certain operations to higher privilege levels. Typically, these privileges form the foundation for how an operating system protects itself from user processes, and protects user processes from each other: the CPU provides high-privileged *kernel* mode and low-privileged *user* mode; to execute sensitive OS operations, the CPU must be in kernel mode; switching from user-mode to kernel-mode requires a special hardware *trap* that also changes execution context to the OS.

Suppose users (including the adversary) can run code in user-mode, and the designers count on the kernel-user system architecture to ensure this code cannot do significant harm. If the adversary can figure out how to cause the CPU to change modes while still executing the adversary's code, then the adversary defeats these protections.

The literature gives us two examples of how, as an unexpected consequence of seemingly innocuous user-level operations, an adversary can do just that. In 1972-1973, as part of a security review of the MULTICS operating system, Paul Karger and Roger Schell discovered that certain carefully laid out instructions on a Honeywell 645 CPU would bypass the memory restrictions set up in kernel mode [KS74, Section 3.2.2]. In 1991, a vulnerability surfaced in Sun SPARCs: an integer division bug could give any user "root" privileges [Neu95, page 116]. It is not clear whether the bug lay in the OS, the hardware, or a combination of the two: however, a software patch fixes it [Sun91].

One wonders at such incidents. Were these design accidents, or deliberate sabotage? Given the difficulty of testing for such features, what can we speculate about the number of other such occurrences in the wild? Does the increased complexity of current CPUs increase the likelihood of such features? How much does the increased sophistication of design verification tools mitigate this risk?

3.5   Erasing Data

When considering memory components, a designer tends to focus on how well these components remember data, and overlook the question of how well the components forget it. For a TCP, this question plays into at least two sets of issues:

- the TCP may use memory that an adversary may later use, and thus should zeroize its contents when done;

- the TCP may defend itself by zeroizing sensitive memory upon tamper.

Peter Gutmann published excellent analyses of how data can *imprint* into magnetic [Gut96] and semiconductor [Gut01] storage. Anderson and Kuhn [AK97] empirically confirmed that most of the key material still remained in an early "secure" cryptographic accelerator even after it had zeroized itself.

My colleague Weingart has experience using radiation and low temperatures to cause CMOS RAM to imprint its contents [Wei00]; as Anderson and Kuhn also observed, an adversary might successfully attack a device that zeroizes itself upon tamper simply by first immersing it in liquid nitrogen. (Here at Dartmouth, where outside temperatures regularly go below $-10°$F in winter, simply leaving the TCP outside for a while may suffice.) Unfortunately, such imprinting data—how long, at what temperature—does not often appear on a device's spec sheet. Voltage spikes are also reputed to cause imprinting.

On the software level, coding to avoid leaving sensitive data in exposed memory is a tricky and often subtle business. Simply having a routine clear its memory before exiting may not suffice, because the compiler may notice that these write operations are never read—and so optimize them away. The operating system may add also act in ways the programmer does not anticipate. For one example, in one embedded OS, a process could explicitly request to share a memory region with another—but the OS will actually share the entire memory page that region contains. For another example, one need only think of what happens to the stack and to the contents of free page frames and free disk sectors, and of deleted data inside a journaling file system.

3.6 System Context

Analyzing a platform for security issues naturally requires focusing on the platform itself. Unfortunately, this focus can sometimes obscure the fact that the platform may only make sense in the context of a larger system for building, deploying, maintaining, and using it. This larger system may contain avenues for an adversary to subvert the platform, without directly attacking the platform.

Example: Software Support for a Secure Coprocessor. Consider a secure coprocessor such as the IBM 4758: significant physical armor, but whose security depends on internal bootstrap and control software, installed and maintained by a privileged entity. At first glance, one focuses on the physical armor; at second glance, one may focus on the software design, its embodiment as source code, and perhaps even its embodiment as the actual executable that runs on the device.

However, the fact that the device may accept signed code updates from the privileged entity brings in a new dimension. What software engineering practices govern the codebase? Is there a good version control system to keep fixed bugs from reappearing? Will changes in the compiler or linking tools affect things? What security practices govern the protection of the private key that signs these updates?

Example: University Smart Cards. For another example, a colleague of mine works at a university where student IDs are chip cards, and students can use these cards to purchase items from vending machines, via a balance stored on the card itself. (The machines also take real money.)

Curious, my colleague's lab built a man-in-the-middle: the chip card plugs in one end, the other end (some plastic, with conductive foil strips emulating the card contacts) plugs into the vending machine, and a handheld computer observes the traffic. Reverse-engineering the card-machine protocol from these observations, these researchers discovered that the card authenticated the vending machine, but not the other way around. (This discovery enabled them to build a handheld-driven simulated card that successfully convinces vending machines to release products.) Furthermore, they discovered that their lab's soft drink vending machine (everything is a same price—let's say $1) would deduct a dollar from the chip card as soon as it was inserted—but then would return the dollar if a purchase was not carried out. When they programmed their simulated card to refuse to accept this dollar back, they discovered that the vending machine—insistent on carrying out its duty—would eject four quarters instead.

Consequently, without directly attacking physical security of the system components (vending machines and chip cards), my colleague was still able to regularly drain his lab's vending machine both of soft drinks and of quarters—and since the party who handled vending machine products differed from the party who collected/replenished vending machine money, no one ever noticed.

Example: University Stored Value Cards. Another colleague tells of his undergraduate university, where students used stored-value cards to purchase photocopies (from photocopy machines) and soft drinks (from vending machines). The photocopy machines offered two different options, priced at (let's say) ten cents and 25 cents. The machines were clever enough to check that card had sufficient stored value for the selected option, when the student inserted the card. However, the machines were not clever enough to check that this value was still sufficient should the student change the selection to the more expensive option after the card was inserted. The machine would attempt to debit 25 cents from a card that contained less, leaving an illegal value stored on the card.

A card in this error state would not subsequently work in a photocopier. It would also not work in a vending machine. However, the vending machine computation did not correctly validate its input; it would try to work with this illegal value anyway. As a consequence, the vending machine would crash (requiring in-person maintenance by appropriate personnel)—but not before writing a legal and significantly large positive value onto the student's card.

As a consequence of this (apparently unforeseen) interaction between three different devices and legal but not necessarily sensible behavior by users, the university regularly found itself in a state where half the vending machines were crashed, the other half were drained of product, and the soft drink supplier was demanding payment.

3.7 Defensive Strategy

Unlike many other researchers in the physical security arena (who focus on attacking systems), Steve Weingart has spent the bulk of his career working on how to build TCPs that can withstand attacks (and then applying this to the physical security architecture of the IBM 4758). The last in the world to claim that a system could be "tamper-proof," Weingart has identified several components that a designer can weave together into a defensive strategy.

3.7.1 Tamper Evidence

The designer can try to ensure that physical attack on a device results in some indelible, observable consequence. Everyday examples include the glass covering on some fire alarms, or the safety-seal on a new bottle of aspirin: the intention is not so much to make tamper difficult as it is to make it obvious to a legitimate user or auditor that tamper has occurred.

Of course, using tamper evidence does not make sense an applications where an appropriate party will not be able to audit the device, or where an after-the-fact observation is too late.

3.7.2 Tamper Resistance

Another technique is to make the device hard to attack physically. A real-world example might be the armor on a bank's safe: the adversary who wants to penetrate this is going to have to obtain significantly powerful tools, and bring them with him to the bank.

3.7.3 Tamper Detection

The designer can include measures to enable the device itself to sense when tamper is occurring. A real-world analogy is a burglar alarm: the adversary may be able to break into the house, but will hopefully end up setting off a motion sensor the homeowner installed.

3.7.4   Tamper Response

Finally, the designer can include measures to enable the device to take some appropriate countermeasure when it detects tamper. Again, in a real-world analogy, a burglar alarm is not effective if detects the burglar, but then rings a very quiet alarm that fails to bring help. In TCP design, tamper response often takes the form of trying to zeroize sensitive data before the adversary can reach it.

3.7.5   Operating Envelope

Our college recently announced migration of our telephone service from traditional lines to voice-over-IP—prompting some wags to ask how one can call the machine room to tell them the network is down.

More seriously, the technology that comprises defense mechanisms itself may require various preconditions and environmental factors to operate correctly. A burglar alarm may require power; tamper response consisting of having the CPU overwrite data requires a functioning CPU for a sufficiently interval; tamper response that focuses on zeroizing SRAM requires the appropriate temperature range in order to avoid imprinting. The designer needs to identify the *operating envelope*—the region within which the device needs to remain for its defense mechanisms to operate correctly—and ensure that the device remains within this region. Applying this principle may require trying to estimate the device's fastest trajectory out of this envelope—and ensuring that the device can successfully complete its defensive countermeasures in time.

3.8   Further Reading

Ross Anderson and his group at Cambridge University have long been leaders in the role physical attacks can play in defeating security architectures. The two Anderson and Kuhn papers mentioned earlier [AK96, AK97] make good reading; Ross's textbook [And01] is also worth a look. Weingart's 2000 paper [Wei00] provides a nice overview of these defense ideas, along with attack avenues. Paul Kocher and his group at Cryptography Research have long been leaders in side-channel analysis [Koc96, KJJ99, for example].

My "Fairy Dust" column [Smi03] in *IEEE Security and Privacy* provides a short survey of these attack issues (and some defense strategies).

Chapter 4

FOUNDATIONS

The bulk of this book focuses on the design and applications of the IBM 4758 secure coprocessor and subsequent TCPA/TCG platforms. However, there are no branches without roots. In this chapter, we'll examine some of the principal ancestors that set the stage for these technologies

Organizing this discussion gives us several choices: we can present things chronologically, or by investigators, or by concept. None of these approaches is "clean," since the concepts are interrelated, and the various projects over time explored various combinations. We'll organize by concept

- Section 4.1 will look at early examinations of applications secure coprocessors, and how they might be integrated in into broader systems.

- Section 4.2 will look at early examinations of physical security and internal architecture for commercially reproducible platforms.

- Section 4.3 will look at using hardware to insure a larger platform boots securely.

- Section 4.4 will look at some relevant projects in the classified/defense world.

4.1   Applications and Integration

If we take as axiomatic that we have some physical protection technology that renders its internal data unavailable to a targeted class of adversaries, the question arises: what do we with it? What security problems do we wish to solve, what computation do we put inside the protected envelope, and how do we position that platform inside a larger system, in order to solve them?

4.1.1   Kent

Stephen Kent's 1980 thesis [Ken80] at MIT systematically explored the use of what he called *tamper-resistant modules (TRMs)* to protect external software. At first glance, this seems like the software piracy problem. In the most straight-forward setting, a software vendor sells one copy of a program to a customer, with the understanding (implicitly or explicitly) that the user only run that copy on one computer; *piracy* consists of making additional copies for use on other computers—and perhaps even resale—without reimbursing the vendor.

However, Kent treats the problem with more generality and foreshadows some of the later work in *mobile agents.* Kent is also concerned with protecting the vendor's interest in preserving integrity and perhaps confidentiality of the software and its operational data, and is also concerned with protecting the user's interests in confining the vendor software to only that part of the user's system environment the user desires.

Kent then develops a taxonomy of system architectures, partitioned mainly into two categories. In the *encrypted bus approach* (generalizing previous work by Best [Bes80]), Kent places the individual elements of the computer system into TRMs, and relies on encryption to protect the communication; in the *encrypted storage approach*, Kent places the processor and some memory inside the TRM, and uses encryption to protect that TRM's interaction with exposed memory. Kent presents a scheme to distribute and authenticate software to the various platforms via a series of shared secret keys.

In many ways, Kent's analysis foreshadowed issues that would later rise to prominence. Externally archiving TCP state could result in replay attacks: since an adversary could "turn back the clock" by restoring an archive [Ken80, p. 46]. The bulk of increased cost of a TCP (over a similar module without physical security) is due to mechanical engineering problems (p. 70). Providing a "maintenance hatch" through which a vendor can physically access the device, for repairs, introduces security weaknesses (pp. 70–71). Encrypting bus traffic still potentially leaves the system vulnerable to traffic analysis (e.g., pp. 106–107)—foreshadowing work in oblivious RAM, In the long run, a secure *virtual machine monitor* may be the best approach to handling mutually suspicious applications (e.g., p. 246).

We will revisit these concepts in later chapters.

4.1.2   Abyss

Steve White and Liam Comerford at IBM Watson then followed up the work by Kent and others, with *ABYSS (A Basic Yorktown Security System)* [WC87]. White and Comerford dismissed the instruction-level protection of an "en-crypted bus" approach as inefficient; instead, they designed the ABYSS system as a TCP that consisted of a small processor and memory, operating in the context of a *host* system that was physically unprotected. In this sense, ABYSS was closer to Kent's encrypted storage architecture, except ABYSS did not necessarily require encrypted storage.

White and Comerford focused on trying to protect software, by ensuring that a customer's use of vendor code would comply with the license restrictions established as part of the purchase. However, rather than trying to secure the entire execution environment of the software, White and Comerford proposed *partitioned* computation: dividing the application into a portion that runs on the exposed host, and a portion that runs inside the TCP. They analyze the semantic and combinatorial complexity of designing a protected portion that an adversary cannot reverse engineer; they also develop a method to distribute this software based on various shared secrets (although they observe that public key cryptography could be used). White and Comerford also develop an extensive system to control and distribute rights-to-execute via smart cards and symmetric cryptography.

(It is interesting to note that White and Comerford indirectly observed that a trustworthy multi-process operating system might be a substantial implementation challenge.)

4.1.3   Citadel

Steve White and colleagues then developed and refined the ABYSS design into a much more comprehensive system, *Citadel* [WWAP91, Pal92].

Rather than trying to secure an entire system by using the TCP to house a *reference monitor*, the Citadel design adopts a *security server* model: the TCP device provides security services to the larger system in which it is embedded. However, rather than confining their vision to ABYSS's view of software protection via partitioned computation, the Citadel designers explored a much more thorough space of potential applications:

- The TCP might function as a *crypto server*, providing cryptographic operations to other elements, but keeping keys and sensitive data within the TCP. The TCP might also be the logical home for specialized cryptographic hardware within the system.

- The TCP might function as an *authentication server*, providing a secure local cache of user authentication and authorization data.

- The TCP might function as a *file/database server*, providing files or responses to clients only when the client is appropriately authorized.

- The TCP might function as an *access control server*, providing resource access to authorized clients only—but otherwise protecting the resource from adversaries that may include the client.

- The TCP might function as an *audit server*, recording logs that cannot be tampered with even by a compromised client.

- The TCP might function as an *execution server*, a generic platform to catch and execute programs in a more secure environment than the client can provide. White et al observe that, in some sense, the execution server is the universal TCP application, since an appropriately designed execution server application can be transformed into instances of any of the others.

The designers also provide a more detailed application scenario, for a hypothetical "Widgco" corporation [WWAP91, pp. 51–54].

White et al consider the many places within a distributed system the Citadel architecture can be places (pp. 38–41), and consider the potential for Citadel units within an IBM PC (pp. 46–48), a mainframe (pp. 48–49), or a smart card (pp. 50–51). Echoing Kent's discussion and foreshadowing later work, the mainframe discussion suggests the potential for embodying the TCP as a secure virtual machine.

As Section 4.1.4 and Section 4.2 below will discuss, Citadel did not remain just a paper design—IBM proceeded to build a number of hardware prototypes.

4.1.4   Dyad

Some of the Citadel prototypes from IBM went to Doug Tygar's group at Carnegie Mellon University, where they became the foundation of the *Dyad* project, built by Ph.D. student Bennet Yee along with Tygar in the early 1990s [Yee94, YT95, TY91, TY93].

Yee and Tygar extended the Citadel vision: although the assumption of physical security is fundamental to the security of distributed systems, "physical security requirements may be isolated to the secure coprocessor." Yee and Tygar went on to demonstrate this explicitly, by developing and implementing an extensive software architecture (as Section 4.2.2 below discusses), and then using this platform to implement several applications:

- Following in the footsteps of Kent and of White and Comerford, Yee and Tygar used the TCP for *copy protection*: the vendor encrypts the protected software so that only the appropriate TCP can decrypt and execute it. Yee even demonstrated the feasibility of White and Comerford's partitioned computation approach by producing a partitioned version of gnu-emacs.

- Following the suggestion of Citadel, Yee and Tygar used a TCP to implement a secure *audit trail*.

- Yee and Tygar used the TCP for fully decentralized *electronic cash*. Here, cash is a balance held inside the TCP. A TCP will only exchange cash when it can recognize another appropriately configured TCP. Implementing cash as a stored value guarded by a computational engine, instead of purely via cryptography [Cha85, for example], enables the use of computation to provide properties the cryptography cannot. For example, standard *transactional* techniques from distributed computation can ensure that the exchange of cash is always atomic, despite network failures. (Yee's work was the first to note that Chaum's DigiCash did not have these properties.) For another example, a computational guard allows cash exchanged to be anonymous, except when they exceed $10K—a requirement of US law.

(It is interesting to note that smart-card-based cash schemes are now common.)

- In both the copy protection and cash applications, Yee and Tygar used the TCPs as a family of protected places that can exchange house and exchange objects, and ensure that use of these objects complies with policies that can travel with the objects. They generalized this to the notion of *electronic contracts*.

- Another application area Yee and Tygar considered was *electronic postage meters*. The US Postal Service permits third-party vendors to lease postal meters to end customers, who use these meters to print indicia on individual pieces of mail. The customers reimburse the meter vendor, who reimburses the postal service. However, adversaries can obtain free postage by copying indicia directly, and by compromising a postal meter itself. Yee and Tygar addressed both problems by designing indicia that incorporates cryptographically signed letter-specific information in a two-dimensional bar code, and then using a Dyad coprocessor to house and wield the secrets that generate this bar code. Thus, the addition of a TCP can transform a personal computer into a postal meter.

- Yee and Tygar also used a TCP to perform a *host integrity check*: to participate in the boot process of the host platform and check each layer of the host software using Karp-Rabin fingerprints. (Using cryptographic techniques is critical: standard error-correcting checksums do not protect against malicious errors; an adversary can modify the software to have his intended altered functionality, but still yield the same checksum.)

This application amplifies the physical security of the TCP to give assurance about the integrity of the larger host platform, although the host security achieved is weaker, since a sufficiently dedicated adversary can subvert it via techniques such as dual-ported memory without breaking the TCP itself. Section 4.3 below considers this family of applications further.

4.2   Architectures

Besides pioneering the use of trusted coprocessors for applications, early work also pioneered the feasibility of actually building real devices.

4.2.1   Physical Security

How can we physically secure a TCP against tamper, in a way that is sufficiently robust and inexpensive to find its way eventually into a mass-produced, deployable technology?

In 1983, Chaum [Cha84] considered the problem, and hinted at the distinctions between tamper-evident design, versus tamper detection and tamper response. Chaum proposed a solution consisting of multiple layers of protective technology, such that no two layers shared vulnerabilities—thus an attack that compromise one layer would be detected by another.   In 1986, Price [Pri86] considered surrounding the device with a hard, tamper-resistant material and embedding conductors in this material to enable tamper detection and response.

However, these discussions were high-level and hypothetical. In 1987, Steve Weingart [Wei87] presented a thorough practical design, for providing physical protection for a device such as an ABYSS card for a PC. Weingart decided the multi-layer approach of Chaum was too complex, and considered a single-layer, embedded conductor approach. Weingart began by considering how *big* a hole the adversary needs to make, in order to mount a useful attack, and decided that 1mm was the lower bound. After experimenting with printed circuit approaches, Weingart decided wrap the package with several layers of fine nichrome wire, and then potting it in a hard material. Winding wires is already a well-understand technology in commercial manufacturing. The winding used multiple strands, to make it easier for the internal tamper-detection circuitry to detect change, and was wound at low tension, to make it harder for the adversary to predict the exact arrangement of wire in any one device. The potting was chemically and physically stronger than the wire, to make it more likely that penetration attempts sufficiently strong to disrupt the potting would also disrupt the wire; the potting also contained alumina or silica, so that attempts to use UV to ablate the potting would generate destructive heat.

Weingart used CMOS technology within the device, to minimize power consumption. His laboratory prototypes of this protective technology could dissipate 2.5W, sufficient for carefully designed circuits. The internal tamper detection circuitry compares the state of the wire winding to a recent sample state (rather than a fixed state), and triggers tamper response if the difference is too great. (Using this sliding window approach permitted the threshold to be much smaller, while also accommodating gradual drift in environmental conditions.) Tamper response disconnects the power of the CMOS RAM, and crowbars its power pin to ground.

In 1987, Weingart noted that CMOS RAM would imprint with low temperatures, but countermeasures were an area of active work.

Weingart's basic design—with nichrome wire replaced by a a conductive mesh printed on a flexible membrane, and with temperature sensing to detect freezing attacks—then saw commercial use in IBM's *Transaction Security System* series of cryptographic accelerators [ADDS91].

4.2.2   Hardware and Software

Citadel. As noted above, the Citadel project at IBM included prototypes. The physical security boundary would enclose a processor (Intel 386, 16Mhz), a megabyte of DRAM, and 64K of battery-backed static RAM that is cleared upon tamper. The prototype also contained a real-time clock. The Citadel design also included an innovative use of FIFO buffers bracketing special DES hardware; this design lets the internal CPU set up cryptographic operations, but then lets the operations proceed without tying up the CPU.

Citadel was designed to have a layered software architecture, with each layer checking the and loading the next, with a focus on using symmetric-key cryptography and on keeping software secret. This process starts with the bootstrap layer. The hardware provides two routes to bootstrap: power-on, and also a reset in order to initiate a code update. However, in order to keep the ROM-resident boot code small and reliable, the design did not permit this bottom layer to use cryptography—which complicates the trust foundation.

The design considered many subtleties. For example, a hardware bootstrap reloads code, because otherwise a malicious code layer could "prevent its own update." For another example, the design worried about maintaining a consistent global state across a distributed network of devices, which may all be undergoing software update.

Dyad. As mentioned above, Yee and Tygar's work included implementing TCP applications on a Citadel prototype. Implementation required establishing a reasonable programming environment within the device. Yee ported the Mach microkernel to run inside Citadel; this task required non-trivial modifications to Mach, in order to accommodate the specialized hardware.

To work around the limited internal memory size, Yee invented and implemented *cryptopaging.* Dyad provided virtual memory to its applications, and used the host as a backing store. Dyad then used Citadel's DES and FIFO hardware to quickly move pages in and out—but encrypting and decrypting them in the process, in order to preserve confidentiality and integrity of page contents from a host-based adversary.

Implementing the applications also required a full instantiation of the layered code-loading process sketched by Citadel. Dyad starts out at boot time with no secrets within the coprocessor, and no internal code other than the primary boot; the other code lives on the host, in suitably protected form. The primary boot code loads the secondary boot code carries an authentication secret with it, for zero-knowledge authentication, as well as cryptographic checksums of the kernel and applications. The Dyad design also noted that the battery-backed secure memory might store secrets that an appropriate administrator may use to upgrade software [Yee94, p. 38]. Some variations require an operator to enter a secret key to decrypt ROM-resident fingerprints [TY91, p. 170]. Using the same basic ideas behind the electronic contract applications, Dyad developed a scheme for securely backing up and migrating state from one coprocessor to another, under appropriate conditions.

Yee's work also included identifying and working around design flaws in the Citadel prototype [Yee94, pp. 42–43]. An interrupt problem hampered throughput. The DES hardware reversed the bits on 2-byte words, and so the FIFO hardware contained "byte flippers" that the software needed to manipulate. More critically, every other decryption "outputs garbage." The prototype he worked with did not have *busmastering* capability: so it could not take over the host bus and probe memory. This prevented a full prototype implementation of the host integrity application.

4.3   Booting

So far, we have discussed early work that developed the applications a that a small trusted computing platform could enable, and that developed the physical realization of these platforms. Concurrently with this work, the related concept of *secure boot* evolved. What can the bootstrap process of a computer system do, to ensure that the system is actually executing the software it is supposed to, in a safe configuration? (See [GGKL89] for an early discussion of boot and loading, in the context of a broader system security architecture.)

Unfortunately, this area can be a bit of a terminology minefield, as some players (e.g., TCPA/TCG—see Chapter 10) use apparently synonymous terms to describe distinct concepts. For example:

- **"Trusted Boot."** Do we want the system to boot the correct configuration?

- **"Secure Boot."** Do we want the system to boot the correct configuration—and halt if it cannot?

- **"Authenticated Boot."** Do we want to be able to verify whatever configuration it was that the system booted?

Since the terminology is not yet standard in the field, we will not be too picky about the usage within this book.

The basic idea behind secure booting is to decompose the system configuration into a series of entities, and have some entity check the integrity (and perhaps identity) of each of these entities. Approaches vary in how they perform this decomposition, how they check integrity, which entity or entities do the checking, and why one should trust them. These approaches suffer from some common potential vulnerabilities:

- Has the entity changed between the time it was checked and the time the system executes it? (This is a classic instantiation of the *time-of-check, time-of-use* problem.)

- Was the initial measurement even of the correct entity? For example, how does a TCP know that the data it just saw really is the operating system an external host was booting?

- How do we assess the correctness of entities that are dynamic? For example, a system may include non-volatile append-only logs that, by definition, will continue to evolve and change through execution and across boots. Simply checking these against a static cryptographic hash will not suffice.

Secure booting is relevant to TCP design for three primary reasons. First, it was one of the first applications developed for TCPs. Secondly, it also can be a technique for building a TCP whose computational boundary exceeds the physical protection boundary. Finally, secure booting is essentially what TCPs with malleable software configurations do within their physical protection boundaries.

Tripwire. In the early 1990 Gene Kim and Gene Spafford at Purdue developed their *Tripwire* system, a software approach to ensuring the integrity of a UNIX platform [KS94]. Tripwire checks that for each file in some selected subset, the contents still match what is considered to be correct value.

Dyad. As mentioned earlier, Yee and Tygar introduced the idea of using an external secure coprocessor to take over the bus of the host during its boot process, and to verify the various software components of the host before letting them boot. They used Karp-Rabin fingerprints, since this approach provided good security and could be implemented quickly; however, they also considered other cryptographic approaches. To help protect against "Trojan hardware" on the host that lies to the coprocessor about what's being measured, Yee suggests having the TCP do random "behavior and timing checks" [Yee94, p.16], but does not elaborate.

BITS. In 1994, Paul Clark and Lance Hoffmann followed up on this idea with their *Boot Integrity Token System (BITS)*, which uses a smart card to assist in checking the integrity of a host operating system. In BITS, the user authenticates to the card via a plaintext password, and then the card and host mutually authenticate via a shared secret. The host reads its boot sector from the smart card; this boot sector then calculates a checksum of the host OS, and compares this with a value stored on the smart card.

Clark and Hoffmann observed that extra processing time at boot is not particularly significant, since users expect booting to take time.

Penn's AEGIS. In 1997, Bill Arbaugh et al at the University of Pennsylvania followed up with AEGIS [AFS97]. Arbaugh starts with a handful of axioms: the initial BIOS can be trusted; a component that has already been checked by a trusted component can be trusted; and no component can be executed until it is trusted. Arbaugh then formally analyzed the boot process of an IBM PC, broke it into a directed acyclic graph, rooted at BIOS, and adjusted BIOS so that each parent checks a cryptographic hash of a child before executing it. An add-on PROM board stores hashes, and provides a facility to fetch a correct version of a software component should the integrity check fail.

Itoi et al [IAPR02] then extended the AEGIS approach to store hashes in a user smart card.

(Unfortunately, MIT started a TCP-related project that also is called "AEGIS," complicating terminology. Chapter 12 discusses this other AEGIS.)

4.4   The Defense Community

The above evolution occurred in the open world of academic and industrial research. However, the US defense and intelligence communities played a major role in shaping the early stages of field of computer security. Some projects relevant to TCPs also emerged from that community.

The *Logical Coprocessing Kernel (LOCK)* was one such project [Say02, for example]. Started in 1973, this project was an attempt to use hardware—along with other tools such as formal assurance processes and capability-based system design, and principals of virtual machine monitors—to build what this subcommunity termed "highly trustworthy" systems. In LOCK, a *System-Independent Domain-Enforcing Assured Reference Monitor (SIDEARM)* implemented an MLS policy; a *Bulk Encryption Device (BED)* encrypted sensitive data as it was stored. Unlike the thrust of TCP work in this book, LOCK seemed to depend on the hardware more for performance than for physical protections. The project died out due to business and engineering reasons.

Another relevant project was the *Fortezza* personal token, used today to house personal keys.

4.5   Further Reading

Bennet Yee's thesis [Yee94] provides a good exposition of all three components: applications, internal architecture, and secure boot; [TYH96] examines the postal meter application in more detail. Weingart's 1987 paper [Wei87] gives good insight into the physical security design. The longer Citadel report [WWAP91] is also good, but does not appear to be available in electronic format. In the secure boot arena, Arbaugh's paper [AFS97] is a nice presentation of the problem.

Chapter 5

DESIGN CHALLENGES

The early experimental work suggested the utility of a a secure coprocessor *platform* as a generic product that enables development and deployment of TCP applications. This chapter discusses the design challenges and tradeoffs we faced, in the mid-1990s, when trying to transform this notion into a real product. Section 5.1 sets the context for this project. Section 5.2 discusses the design obstacles we faced. Section 5.3 boils down this analysis into a set of requirements for a product. Section 5.4 sketches the technology we developed to help overcome these obstacles, and some usage scenarios this technology might enable.

5.1 Context

From the ABYSS and Citadel work, the notion of a secure coprocessor emerged. This work, along with Weingart's work on physical security, demonstrated that building secure coprocessor platforms in quantities sufficiently large to be widely deployed might be feasible. The Dyad work demonstrated that using secure coprocessors to solve real security problems in distributed environments might be feasible. This progression, combined with a number of other threads, led to the my arrival at IBM Watson in 1996, and subsequent participation in creating a high-end secure coprocessor platform as a commercial product.

Two paths—personal and commercial—led to this situation.

5.1.1 Personal

On a personal level, in the mid-1990s, I was working in the Computer Research and Applications group at Los Alamos National Laboratory. At that point—the dawn of the Web—many enterprises were migrating their legacy services—based on paper, telephone, and offices—to the new electronic environment the Web offered. As the nature of "distributed computing" fundamentally changed, I was part of a team doing vulnerability analyses and security designs for clients, primarily public-sector, considering this migration.

The application and deployment scenarios we were encountering were rife with situations where one party needed to trust the integrity and occasionally confidentiality of computation taking place on a machine controlled by a different party with different interests. Being part of Tygar's group at CMU had exposed me to Citadel and Dyad, and also convinced me of the potential for secure coprocessing to address the trust scenarios emerging in this new information infrastructure (e.g., see [Smi96]).

My arguments convinced my superiors to fund a research project in secure coprocessor applications. However, this funding was contingent on several criteria. I had to build the applications on real secure coprocessors, not lab toys, and I needed to be able to scale up gradually through a series of pilots: ten units, then 100 units, then a thousand units. Unfortunately, the platform necessary for me to follow these constraints did not exist. Citadel remained a small number of hand-built prototypes ("about half of which work," one engineer quipped); smaller PCMCIA-based tokens available at the time provided much weaker security and programming environments, and vendors did not want to make them available to me unless I could demonstrate a market for tens of thousands of units—in which case they might "help" me develop my applications.

Elaine Palmer, leading the Citadel project at IBM, suggested instead that I come to IBM Watson and do secure coprocessor research there. I accepted.

5.1.2 Commercial

Without real platforms in large quantities, secure coprocessing would remain an academic exercise. Production of such platforms would require significant investment of resources; except perhaps for the brief dot-com bubble, such investment does not occur without specific business objectives.

When I arrived at IBM and signed the employee confidentiality agreement, the situation was explained. The business arm of IBM had long supported the *Common Cryptographic Architecture (CCA)*, an API for cryptographic services. By the mid-1990s, IBM was marketing a hardware cryptographic accelerator for CCA, and realized that its next-generation accelerator would require many of the properties that the ABYSS and Citadel (and Dyad) researchers had been proposing for secure coprocessors. This realization led to a unique opportunity: the Watson researchers who had been preaching the secure coprocessor vision suddenly had funding to define and design the sought-after generic platform—and to help build it as a real product. But this opportunity was coupled with a Faustian twist: the platform must satisfactorily support an application that implements the CCA API. (This design constraint became significantly more burdensome with Mike Bond's later discovery [BA01] of flaws in this version of CCA API—which the many in the media then transformed to "4758 flaws". Even as we go to press, another cryptanalysis paper has surfaced [PH04] that misses the fundamental difference between the 4758 platform and the CCA application.)

This was the challenge: define and build the trusted computing platform we always wanted—but bring it in on time, and make sure that it can be turned into a crypto accelerator that satisfies IBM's business needs.

5.2 Obstacles

Realizing this vision of high-end secure coprocessors as trusted computing platforms requires overcoming the significant obstacles that confront an organization wishing to develop and deploy real applications using this technology.

5.2.1 Hardware

Tamper Protection. As we have discussed, most discussions of secure hardware usually use phrases like "tamper-proof," "tamper-resistant," "tamper-evident" or "tamper-responsive." Any realistic assessment recognizes that "tamper-proof" hardware is unattainable. However, our basic model assumes that the TCP somehow renders some stored data unavailable to some adversaries. In stronger platforms, these adversaries may have direct physical access, and thus one wants to sketch these TCPs as somehow responding to tamper by zeroizing sensitive information.

How should this happen? Initially, we can divide the approaches into *active* and *passive*. Active tamper-response relies on the device itself to detect tamper attempts and destroy its secrets. Active response can be computational—which requires that during tamper, the processor remain alive long enough to destroy the secrets—or depend instead on independent special-purpose circuitry that more quickly crowbars the memory. Passive techniques rely on physical or chemical hardness (and sometimes on explosives).

Passive protection is difficult to carry out effectively (witness the continued permeation of smart card technology) and difficult to apply to multi-chip modules. But on the other hand, using active protection requires recognizing that the device is only as secure as long as the necessary environment exists for the active protection to function. Minimally, this recognition requires grappling with some difficult issues:

- The continuous existence of this environment requires a continuous source of power.

- What exactly do we know about the device after an interval in which this environment fails to exist?

- Does the device always protect itself, or only between visits by a security officer?

- What should happen to a device that zeroizes its secrets?

Exactly how the zeroizable secrets should be stored raises additional design issues—for example, as Section 3.5 discussed, various environmental conditions can cause static RAM to become imprinted with the data that is supposed to be erased.

Application-specific contexts can suggest other avenues for tamper protection. For example, Anderson and Bezuidenhoudt discuss an electric power meter that used *social pressure* as a tamper-protection technique: if it detected tamper, it responded by shorting out power, tripping a local breaker and blacking out power for local users for over a day [AB96].

Trusted Paths.   Effectively using a trusted device in human-based applications often requires effective authentication of communications between a human and their trusted device [GSTY96]. A human-usable I/O path on the secure device itself makes these problems simpler—but although a nice abstraction, such a path can greatly compromise the physical security, since we need to punch a hole through the armor to let the LCD or keypads through.

Hosts.   A TCP that exists as a secure coprocessor requires, almost tautologically, a host system. Wide deployment of a secure coprocessor application requires considering the population of host machines:

- What physical interface should be used? How does this choice affect ease of installation, number of potential platforms, and performance of coprocessor?

For example, chip-card, PCI-bus, and USB interfaces all give different answers to these questions. For a time in the mid-1990s, the PC-card (at that time, "PCMCIA") interface and form factor seemed poised to become the vehicle for secure user tokens, because PC cards were portable and PC slots were becoming ubiquitous. At the time, however, one vendor told me how the connector on the test rig at the end of their manufacturing process needed to be replaced regularly, because the connectors were only designed for limited number of insertions/removals. Had we deployed an application where a large number of users used a PC card to authenticate to some public machine, this limited lifetime might have caused problems—user cards would have been fine, but the connectors on multi-user machines would have failed.

- What device drivers and other associated host-side software are required? How does this software get to the host? What possibilities exist for attacking the application by attacking the host-side software?

- How much does the coprocessor depend on the host for functions such as storage of code or encrypted virtual memory pages? Does the coprocessor "outsource" actual computation to the host, perhaps after performing some type of integrity check?

Cost and Durability. For an application to succeed, someone needs to create and distribute a population of secure coprocessors. This task requires balancing the cost of the device with its power and protections, as well as considering longer-term reliability issues. (Indeed, the often-lamented computational restrictions of chip cards are a consequence of the requirement to keep them highly robust to physical wear-and-tear.)

Exportability. For a long time, another challenge facing any practical development and deployment of cryptographically powerful devices was compliance with the U.S. export laws, which limited what capabilities could be included in technology shipped outside the U.S. Regarding this topic, one typically hears political and idealogical discussions: ranging from assertions that the true purpose was to suppress strong cryptography in domestic products, in order to facilitate illegal domestic spying, to assertions that it is critical to keep strong cryptography away from terrorists.

From a designer's perspective, the consequence of the export laws was not ideology but complexity. It was not a simple matter of "non-U.S. $\Rightarrow$ weak crypto." Rather, various customers, lines of business, and regions might have special permissions for certain types of operations, and we needed to take into account this rather complex (and potentially dynamic) policy space when designing generic boxes.

5.2.2 Software

Using a TCP requires putting application-specific software on it. Developing and deploying TCP applications thus requires the ability to develop software for the device. This requirement led to many challenges:

- Is development possible on a *small scale* with small numbers of devices—or must the application developer first convince a hardware manufacturer of the business case for thousands or millions of units?

- Is development possible *independent* of the hardware manufacturer—or must the application developer work closely and share plans and code with the manufacturer?

- Does a robust *programming environment* exist for the device, or must code be hand-tuned? What about debugging and testing?

- If independent development is possible, what prevents malicious or faulty software from *compromising* core device keys? Do these protections consist of verified hardware and software, or depend solely on complex software with a track record of flaws?

Once application software is developed, the developer needs to ensure that the potentially untrusted user, in a potentially hostile environment, ends up with an authentic, untampered device that is programmed with the right software.

The developer could install the software at the factory. However, this option forces the application developer to have a substantial presence in the factory, and forces the factory to customize their processes to individual application developers. This approach may be prohibitively costly for small-scale development. This option also complicates the manufacturing process: each new variation on a shippable product explodes the amount of paperwork and bookkeeping, particularly for well-established vendors.

However, installing software at some later point, after the TCP leaves the factory, raises additional issues.

- What about the security of the shipping channel? What if the device is modified between the time it leaves the factory and the time the software is installed? If, as part of a vulnerability analysis, one considers how to maximize the gain from a $20K budget, bribing a truck driver might be rather effective. (I have seen at least one well-respected speaker overlook the fact that if a vendor can package a real device in a way that does not look tampered, then perhaps an adversary with access to the shipping channel could as well.)

- How does the device know what software to accept? (Accepting just anything opens the possibility of tamper via false software load.)

- Does a device carry a key—or secret software—whose exposure compromises that device, or other devices?

- With general-purpose programmable hardware intended for multiple application developers, installation after the factory needs to ensure that hardware loaded with one developer's software cannot claim to be executing software from a different developer.

If not at the factory, the application developer might install software at their own site, or at the end-user site. Installing the software at the application developer forces the developer to ship the hardware to the end-users. Installing the software at the end-user's site requires the need for security officers, or for the device itself to exert fairly robust control, authentication, and confirmation of software loads.

The software installation process may also have unpleasant interactions with the desired security model. For example, if the application developer requires that their software itself be secret, but the hardware only provides a limited amount of tamper-protected storage, then the installation process must include some way of installing the software decryption key in that storage.

Most post-factory installation scenarios require that the devices leave the factory with some type of security/bootstrap code, which raises additional issues.

Software Maintenance. After installation, how do the software developers then proceed to securely carry out the maintenance and upgrades that such complex software inevitably requires?

- How does the device authenticate such requests? Must the developer use an on-site "security officer," or can they use remote control? If the latter, how much interaction is required? Does the application developer have to undergo a lengthy handshake with each deployed device? Does the developer need to maintain a database of device-specific records or secrets?

- How can participants in an application know for certain that an upgrade has occurred? (After all, the purpose of the upgrade might be to eliminate a software vulnerability which an adversary has already used to explore the contents of privileged memory.)

- What should happen to stored data when software is upgraded? Not supporting "hot updates" that preserve secrets is cleaner, but can greatly complicate the difficulty of performing updates. (Some application developers in the financial cryptography arena have insisted that all updates destroy all secrets, because that makes it easier to reason about security. Others say that customers complain when they need to back-up and reload keys due to a routine software upgrade they already trust.)

- What atomicity does the device provide for software updates? Can failures (or malice) leave the device in a dangerous or inoperable state? What if the software that cryptographically verifies updates is itself being updated?

To avoid grappling with these issues, some developers may choose simply to not allow updates. However, the decision certainly needs to be balanced against hardware expense and software complexity (hence likelihood of upgrade).

Multi-Party Issues. The foregoing discussion largely focused on a model where basic device hardware had one software component that needed to be installed and updated. In reality, this situation may be more complicated. Multiple software layers may lie beneath the application software:

- The presence of a device operating system (in order to make software development easier) raises the questions of when, where, and how the OS is installed and updated.

- The OS may come from an independent software developer, like the application does. This scenarios gives rise to potentially four different entities involved in a TCP configuration: the hardware manufacturer, the OS developer, the application developer, and the end-user.

- The more tasks assigned to the basic bootstrap/configuration, the more likely this foundational software might also require update.

Multiple layers each controlled by a different authority makes the software installation and update problem even more interesting. For example, the ability to perform "hot-updates" potentially gives an OS vendor a backdoor into the application secrets. What if the application developer does not necessarily trust the OS vendor to be honest—or to release bug-free updates? What if the application developer wants to reserve the right to inspect and approve updates before letting them come in under the application?

The simple answer of "not allowing any OS updates" avoids these risks, but introduces the problem of what to do when a flaw is discovered in security-critical system software—especially if this software is too complex to have been formally verified. (Formal verification does not eliminate the potential of flaws, but would at least provide greater assurance of their absence.)

Some scenarios may additionally require multiple sibling software components, at the same layer (although this flexibility must be balanced against the risks of potentially malicious sibling applications, the hardware expense and the sensitivity of the application).

Comparison to PCs. It might be enlightening to compare this situation with software development for ordinary, exposed machines, such as personal computers. For PCs, software developers do not need to build and distribute the computers themselves. Software developers never need meet or verify the identity of the user. Software developers do not need to worry about how or where the user obtained the machine; whether it is a genuine or modified machine, or whether the software or its execution is being somehow modified. Furthermore, developers of the application software usually do not also have to develop and maintain the operating system or the ROM BIOS.

How do we preserve these features of the legacy software market in the TCP arena? Should we?

The list of obstacles to deploying TCP applications naturally leads to design issues for those hoping to minimize these obstacles by building such platforms.

Lessons Learned. The issues here arose in part from the collective experience of the IBM research group tasked with developing a TCP—and our connections to the business units who had been selling physically hardened crypto accelerators. Our design choices for this product were driven not just by academic analysis, but also by more practical factors and lessons learned. For example:

- Software is less stable than hardware—especially if the time delay between manufacture and end-user installation is considerable.

- The complexity of manufacturing and maintenance support appears to increase exponentially with each shippable variation of a commercial product.

- No one wants to trust anyone else more than necessary. The end-user does not really want to trust any of the software vendors; the software vendors do not trust each other; and everyone is suspicious of the manufacturer.

- Expensive hardware must be repairable, if possible.

5.3 Requirements

Section 5.2 sketched some design obstacles. We now elaborate these thoughts into a more formal consideration of requirements that drive the architecture.

In order to be effective, our solution must simultaneously fulfill two different sets of goals. The device must provide the core security and trust properties necessary for TCP applications. But the device must also be a practical, commercial product; this goal gives rise to many additional constraints, which can interact with the security properties in subtle ways.

5.3.1 Commercial Requirements

Our device must exist as a programmable, general-purpose computer.

To begin with, the goal of supporting the widespread development and deployment of applications has many implications:

- The device must be easily programmable.

- The device must have a general-purpose operating system, with debugging support when appropriate.

- The device must support a large population of authorities developing and releasing application and OS code, deployed in various combinations on different instantiations of the same basic device.

- The device must support vertical partitioning: an application may come from one vendor, an OS from another, bootstrap code from a third.

- These vendors may not necessarily trust each other—hence, the architecture should permit no *backdoors*: ways for vendor Alice to gain access to vendor Bob's data or code, in a way that Bob did not approve.

The process of manufacturing and distribution must be as simple as possible:

- We need to minimize the number of variations of the device.

- It must be possible to configure the software on the device after shipment, in what we must regard as a hostile environment.

- We must reduce or eliminate the need to store a large database of records— secret or otherwise—pertaining to individual devices.

- As an international corporation based in the United States, we must abide by US export regulations.

The complexity of the proposed software—and the cost of a high-end device— mean that it must be possible to update the software already installed in a device.

- These updates should be safe, easy, and minimize disruption of device operation.

- When possible, the updates should be performed remotely, in the hostile field, without requiring the presence of a trusted security officer.

- When reasonable, internal application state should persist across updates.

- Particular versions of software may be so defective as to be non-functional or downright malicious. Safe, easy updates must be possible even then.

- Due to its complexity and ever-evolving nature, the code supporting high-end cryptography—including public key, hashing, and randomness—must itself be updatable. But repair should be possible even if this software is non-functional.

The reader should note that the design choices often interact in subtle ways. For just one example, the business decisions to support remote update of potentially buggy kernel-level software requires the ability to remotely authenticate that this repair took place, which in turn may require changing the hardware to provide a region of secure memory that is private even from a defective supervisor-level operating system.

5.3.2   Security Requirements

The primary value of a TCP is its ability to provide a trustworthy and trustable sanctuary in a hostile environment. This goal leads to two core security requirements:

- The device must really provide a safe haven for application software to execute and accumulate secrets.

- It must be possible to remotely distinguish between a message from a genuine application on an untampered device, and a message from a clever adversary.

We consider these requirements in turn.

Safe Execution. It must be possible for the card, placed in a hostile environment, to distinguish between genuine software updates from the appropriate trusted sources, and attacks from a clever adversary. The foundation of TCP applications is that the platform really provides safe haven. For example, suppose that, like Yee, we are implementing decentralized electronic cash by having two secure devices shake hands and then transactionally exchange money. Such a cash program may store two critical parameters in tamper-protected memory: the private key of this wallet, and the current balance of this wallet. Minimally, it must be the case that physical attack really destroys the private key. However, it must also be the case that the stored balance never change except through appropriate action of the cash program. For example, the balance should not change due to defective memory management or lack of fault-tolerance in updates. Formalizing this requirement brings out many subtleties, especially in light of the flexible shipment, loading, and update scenarios required above. For example:

- What if the adversary physically modifies the device before the cash program was installed?

- What if the adversary "updates" the cash program with a malicious version?

- What if the adversary updates the operating system underneath the cash program with a malicious version?

- What if the adversary already updated the operating system with a malicious version before the cash program was installed?

- What if the adversary replaced the public key cryptography code with one that provides backdoors?

- What if a sibling application finds and exploits a flaw in the protections provided by the underlying operating system?

After much consideration, we developed safe execution criteria that address the authority in charge of a particular software layer, and the execution environment— the code and hardware—that has access to the secrets belonging to that layer.

- **Control of software.** If Authority N has ownership of a particular software layer in a particular device, then only Authority N, or a designated superior, can load code into that layer in that device.

- **Access to secrets.** The secrets belonging to this layer are accessible only by code that Authority N trusts, executing on hardware that Authority N trusts, in the appropriate context.

5.3.3   Authenticated Execution

Providing a safe haven for code to run does not do much good, if it is not possible to distinguish this safe haven from an impostor. It must thus be possible to:

- authenticate an untampered device;

- authenticate its software configuration; and

- do this remotely, via computational means.

The first requirement is the most natural. Consider again example of decentralized cash. An adversary who runs this application on an exposed computer but convinces the world it is really running on a secure device has compromised the entire cash system—since he or she can freely counterfeit money by incrementing the stored balance.

The second requirement—authenticating the software configuration—is often overlooked but equally important. In the cash example, running a maliciously modified wallet application on a secure device also gives an adversary the ability to counterfeit money. For another example, running a Certificate Authority on a physically secure machine without knowing for certain what key generation software is really installed leaves one open to attack [YY96].

The third requirement—remote verification—is driven by two main concerns. First, in the most general distributed application scenarios, participants may be separated by great physical distance, and have no trusted witnesses at each other's site. Physical inspection is not possible, and even the strongest tamper-evidence technology is not effective without a good audit procedure. Furthermore, we are reluctant to trust the effectiveness of commercially feasible tamper-evidence technology against the dedicated adversaries that might target a high-end device. Tamper-evidence technology only attempts to ensure that tampering leaves clear visual signs. We are afraid that a device that is opened, modified and reassembled may appear perfect enough to fool even trained analysts.

This potential for perfect reassembly raises the serious possibility of attack during distribution and configuration. In many deployment scenarios, no one will have both the skills and the motivation to detect physical tamper. The user who takes the device out of its shipping carton will probably not have the ability to carry out the forensic physical analysis necessary to detect a sophisticated attack with high assurance. Furthermore, the user may be the adversary—who probably should not be trusted to report whether or not his or her device shows signs of the physical attack he or she just attempted. (Again, consider who benefits from tampering with an electronic wallet—or a postal meter.) Those parties—such as, perhaps, the manufacturer—with both the skills and the motivation to detect tamper may be reluctant to accept the potential liability of a *false negative* tamper evaluation. For all these reasons, our tamper-protection approach must not rely on tamper-evidence alone.

Computational Power. A question that can arise in a design decision is how much computational power and memory should live within the protected environment of a TCP. On an engineering level, the answer may rest on issues such as how much heat dissipation the physical packaging allows, what type of memory can be effectively zeroized, and what commercial chip packaging is available. On an economic level, the answer may rest on the "business case": how much the customer is willing to pay for, and (to a lesser extent, perhaps) how reliable the technology appears to be. (For example, should a designer put a hard disk inside the packaging?) On a security level, we might think about how well-funded the adversary is likely to be, and how long the security of the TCP should last. (For example, designing a box that will remain tamper-protected even against adversaries whose technologies are 20 years ahead of today might be rather tricky.)

However, this rather simple question belies some deeper issues. How much computation and memory is enough to secure a broader application? In the mid-1990s, a JavaCard pioneer chided me that "you guys are always saying 'oh, if only we had a little more CPU.' " More serious reflection sees some alternate computational models emerging: with a magic box of size N that has a certain list of features, how big a problem can be "secured?" We'll revisit this issue in Chapter 9.

5.4   Technology Decisions

We decided on building a board-level coprocessor assembled, as much as possible, from existing commercial technology. For protection, we chose active tamper response (electrical, not computational). In an attempt to broaden the family of compatible hosts, we used a PCI interface and developed host-side software for several popular operating systems.

Figure 5.1 sketches the hardware architecture of the coprocessor: a 486-class CPU; accelerators for modular exponentiation and DES (and, in the follow-on, TDES and SHA-1); noise-based random number generation. Volatile dynamic RAM provides the main operational store; battery-backed static RAM (*BBRAM*) provides the non-volatile protected memory. FLASH provides long-term storage for unprotected data. One sector of the FLASH is hardwired as ROM (as Section 3.4.2 discussed).

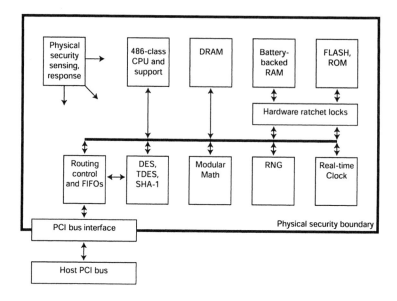

Figure 5.1. The basic hardware architecture.

Making a commercial product support our TCP application design requires giving the device a robust programming environment, and making it easy for developers to use this environment—even if they do not necessarily trust IBM or each other.

These goals led to a multi-layer software architecture:

- a foundational *Miniboot* layer manages security and configuration;

- an *operating system* layer manages computational, storage, and crypto-graphic resources; and

- an unprivileged *application* layer uses these resources to provide services

("Miniboot" was named because it needed to be simple enough to be secure with high assurance, and because it needed to run at boot time.)

Figure 5.2 sketches this architecture. Miniboot consisted of two components: Miniboot 0, residing in ROM, and Miniboot 1, which resides, like the OS and the application, in rewritable non-volatile FLASH memory.

To simplify the production process and comply with export laws, all devices were shipped the same: with only the bootstrap layer. Software installation (and update) could occur at any point thereafter, including at the end-user site, via broadcast-style commands from remote authorities.

In practice, IBM made available two optional code-load sequences that came "for free" for a coprocessor:

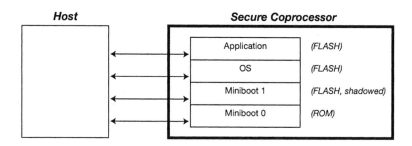

Figure 5.2. The basic software architecture.

- The *CP/Q++* operating system layer consists of the CP/Q embedded OS, pared down to eliminate modules unnecessary for this hardware and enhanced with "managers" providing interfaces to the cryptographic hardware and BBRAM.

- The *CCA* application layer lives on top of CP/Q++, and turns the box into a crypto accelerator.

To enable development, IBM provided code-load commands that allowed a user to switch between standard CP/Q++ and a debug-enhanced one, and to install arbitrary applications in the application layer.

Our device was not a portable user token. Because of our choice of active tamper response, factors such as low temperatures, x-rays, and bungled battery changes may all trigger zeroization—since otherwise, these are avenues for undetectable tamper. (Chapter 6 discusses the technical details of this architecture.)

Usage Scenarios. Enabling widespread development and deployment of secure coprocessing applications required a tool that was easily programmable.

However, adding sufficient computational power and physical security resulted in a TCP sufficiently expensive that the ability to update software became necessary. The fact that our device is more a fixed extension of the host than a highly portable user token made supporting field installation of software a necessity.

Essentially, we converged on the generic PC model discussed in Section 5.2, and attempted to maximize independence between the development/distribution of the hardware, and the development/distribution of the software. As with PCs, we wanted end users of our TCP to obtain their hardware from anywhere, and install the software on their own. But unlike PCs, bona fide software installed into an untampered device can authenticate itself as such.

Designing for this "worst-case" approach—software from multiple parties gets installed and updated in the hostile field, without security officers—permits a wide range of development, deployment, and usage scenarios.

Application Development.

- This design could support *off-the-shelf applications*. A party wishing to deploy an application may find that suitable software (for example, application-layer code that transforms the device into a crypto accelerator providing whatever crypto API and algorithms are currently fashionable) is already available.

- This design could support *off-the-shelf operating systems*. A party wishing to deploy a more customized application may choose an OS that is already available, register with that vendor, and build on that programming environment.

- This design could support *debug and development*. Debug and development environments become just another variation of the operating system—the developer can use an off-the-shelf device, with a different OS load.

- This design could support *on the metal* applications. A party wishing complete on-the-metal control of the device can register with the manufacture, and take control of the OS layer in new off-the-shelf devices.

- This design could also support *incremental validation*. Stakeholders in a TCP application deployment might want more assurance that they can trust in this platform than mere vendor claims can provide. However, existing validation programs—such as (then) FIPS 140-1 [Nat94]—apply to an entire configured device, and can be extremely painful to undergo. Secure field upgradability would make our makes a device not a "FIPS 140-1 Module" per se but rather a partially certified "meta-module." If the hardware and bootstrap software have FIPS validation, a developer need only submit their additional code to obtain a fully validated module for their application. (Chapter 8 considers this topic further.)

Application Deployment.

- This design can support software distribution by *remote broadcast*. A developer can avoid the hassle of distributing hardware themselves by just registering as a code vendor and publishing a download command on the Web. The end-users can purchase the hardware from any standard manufacturer channel.

■ This design can support software distribution by *remote handshake*. If the developer would like more control, he can use targeting and authentication features of the device bootstrap to interactively install software into a particular device, remotely over an open network. This installation could even involve encrypting the software for that device—hearkening back to the designs of Citadel, Yee, Tygar, and Kent.

■ This design can support software distribution by *local security officers*. The developer can always eliminate the open network, and send an authorized security officer with their own trusted device to perform installation.

■ This design can support software distribution by *direct shipment*. A developer can also follow the traditional model of obtaining the devices, install the software, and ship them to their users.

Our long-term vision was commerce and computation among people who have never met. The ability for devices to authenticate themselves and their software configurations permits users of such secure devices to securely interact with each other remotely, across an open network—even if these users have met neither each other nor the application developer.

5.5 Further Reading

Section 5.2 was based on Section 4 from my 1998 *Financial Cryptography* paper [SPW98], and Section 5.4 was based on Section 5 and Section 6. Section 5.3 was based on Section 2 from my 1999 *Computer Networks* paper [SW99].

[DPSL99] provides more information on the internal software design of CP/Q++; [DLP⁺ 01] provides a retrospective looking at the product effort.

Chapter 6

PLATFORM ARCHITECTURE

Chapter 2 laid out some motivations for TCPs. Chapter 3 surveyed the attack space. Chapter 4 reviewed some early design work in this area. Chapter 5 set the stage that resulted: my group at IBM had the chance to design and build a generic secure coprocessor platform, as a product, to enable TCP applications in the real world (even though IBM thought they were getting a crypto accelerator); however, this design needed to satisfy a range of commercial and security constraints.

This chapter lays out the the security architecture I developed with Steve Weingart to address these problems.

One of the lessons I learned from this design experience is that elements of the design cannot be considered in isolation from each other. Consequently, this chapter begins by discussing the overall security architecture that we developed (Section 6.1). It then introduces each individual component: ensuring that secrets are destroyed upon tamper (Section 6.2); ensuring that secrets start out secret (Section 6.3); ensuring that the flaws inevitable in a rich computational environment do not reveal these secrets (Section 6.4, Section 6.5); and enabling developers to develop, deploy, and maintain code (Section 6.6). Section 6.7 then sketches how all these pieces work together.

(Later, Chapter 7 will discuss how we ensure the resulting secure coprocessor application can prove it is "the real thing, doing the right thing"; Chapter 8 will discuss the formal modeling and validation techniques we used to increase assurance that the design works.)

6.1 Overview

In order to meet the requirements of Chapter 5, our architecture must ensure secure loading and execution of code, while also accommodating the flexibility and trust scenarios dictated by commercial constraints.

6.1.1   Security Architecture

Secrets. Discussions of secure coprocessor technology usually begin with "physical attack zeroizes secrets." Our security architecture must begin by ensuring that tamper actually destroys secrets that actually meant something. We do this with three main techniques:

- **The secrets go away with physical attack.** Section 6.2 presents our tamper-detection circuitry and protocol techniques. These ensure that physical attack results in the actual zeroization of sensitive memory.

- **The secrets started out secret.** Section 6.3 presents our factory initialization and regeneration/recertification protocols. These ensure that the secrets, when first established, were neither known nor predictable outside the card, and do not require assumptions of indefinite security of any given key pair.

- **The secrets stayed secret despite software attack.** Section 6.4 presents our hardware ratchet lock techniques. These techniques ensure that, despite arbitrarily bad compromise of rewritable software, sufficiently many secrets remain to enable recovery of the device.

Code. Second, we must ensure that code is loaded and updated in a safe way. Discussions of code-downloading usually begin with "just sign the code." However, focusing on code-signing alone neglects several additional subtleties that this security architecture must address. Further complications arise from the commercial requirement that this architecture accommodate a pool of mutually suspicious developers, who produce code that is loaded and updated in the hostile field, with no trusted couriers.

For code loading and update, we must have techniques that address questions such as:

- What about updates to the code that checks the signatures for updates?

- Against whose public key should we check the signature?

- Does code end up installed in the correct place?

- What happens when another authority updates a layer on which one's code depends?

For the code loading techniques to be effective, we must also address issues such as:

- What about the integrity of the code that checks the signature?

- Can adversarial code rewrite other layers?

Section 6.5 presents our techniques for code integrity, and Section 6.6 presents our protocols for code loading. Together, these ensure that the code in a layer is changed and executed only in an environment trusted by the appropriate code authority.

Goals. Our full architecture carefully combines these building blocks described to achieve the required security properties.

- Section 6.7 presents how our secrecy management and code integrity techniques interact to achieve the requirement that software loaded onto the card can execute and accumulate state in a continuously trusted environment, despite the risks introduced by dependency on underlying software controlled by a potentially hostile authority.

- Chapter 7 will present how our secrecy management and code integrity techniques interact to achieve the requirement that a relying party can distinguish between a message from a particular program in a particular configuration of an untampered device, and a message from a clever adversary.

6.2 Erasing Secrets

The physical component of the security architecture was designed by Steve Weingart, and continued in the tradition of his μABYSS work (Section 4.2.1) and his physical defense design philosophy (Section 3.7).

The main goal of physical security is to ensure that the hardware can detect if it remains in an unmolested state—and if so, to ensure that the hardware continues to work in the way it was intended to work. To achieve physical security, we started with our basic computational/crypto device and added additional circuitry and components to detect tampering by direct physical penetration or by unusual operating conditions. If the circuit detects a condition that would compromise correct operation, the circuit responds in a manner to prevent theft of secrets or misuse of the secure coprocessor.

We felt that commercially feasible tamper-evidence technology and tamper-resistance technology cannot withstand the dedicated attacks that a high performance, multi-chip coprocessor might face. Consequently, our design incorporates an interleaving of resistance and detection/response techniques, so that penetrations are sufficiently difficult to trigger device response.

Historically, work in this area placed the largest effort on physical penetration. Preventing an adversary from penetrating the secure coprocessor and probing the circuit to discover the contained secrets is still the first step in a physical security design. As feasible tampering attacks become more sophisticated through time and practice, it has become necessary to improve all aspects of a physical security system. Designs get better and better, but so do the adversary's skill and tools. As a result, physical security is, and will remain, a race between the defender and the attacker.

(Fortunately, to date, we are not aware of a successful physical attack on what we ended up building here.)

6.2.1   Penetration Resistance and Detection

Building on the earlier μABYSS work, we use a grid of conductors monitored by circuitry that can detect changes in the properties (open, shorts, changes in conductivity) of the conductors. The conductors themselves are non-metallic and closely resemble the material in which they are embedded—making discovery, isolation, and manipulation more difficult. We arrange these grids in several layers; the sensing circuitry can detect accidental connection between layers as well as changes in an individual layer.

The sensing grids were made of flexible material and are wrapped around and attached to the secure coprocessor package as if it were being gift-wrapped. Connections to and from the secure coprocessor were made via a thin flexible cable which is brought out between the folds in the sensing grids so that no openings were left in the package. (Using a standard connector would leave such openings.) After we wrapped the package, we embedded it in a potting material. As mentioned above, this material closely resembles the material of the conductors in the sensing grids. Besides making it harder to find the conductors, this physical and chemical resemblance makes it nearly impossible for an attacker to penetrate the potting without also affecting the conductors. Then we enclosed the entire package in a grounded shield to reduce susceptibility to electromagnetic interference and to reduce detectable electromagnetic emanations.

6.2.2   Tamper Response

Upon detection of tamper, we zeroize the BBRAM and disable the rest of the device by holding it in reset. The tamper detection/response circuitry is active at all times, whether the coprocessor is powered or not—the detection/response circuitry runs on the same battery that maintains the BBRAM when the unit is unpowered.

Tamper can happen quickly. In order to erase quickly, we crowbar the SRAM by switching its power connection to ground. At the same time, we force all data, address and control lines to a high impedance state, in order to prevent back-powering of the SRAM via those lines. (One engineer tells a story of debugging a motherboard that worked almost, but not quite, correctly. It turns out that the main CPU's power pin was disconnected—but the CPU was deriving enough power from its control lines to operate.)

We employ this technique — crowbar and tri-state — because it is simple, effective, and it does not depend on the CPU being sufficiently operational for sufficiently long to overwrite the contents of the SRAM on tamper.

6.2.3   Other Physical Attacks

To prevent attacks based on manipulating the operating conditions, including those that would make it difficult to respond to tamper and erase the secrets in BBRAM, we added several additional sensors to the security circuitry to detect and respond to changes in operating conditions.

As Section 3.5 discussed, for zeroization to be effective, certain environmental conditions must be met. To prevent imprinting via low temperatures, a temperature sensor in our device will cause the protection circuit to erase the BBRAM if the temperature goes below a pre-set level. To prevent imprinting and circuit disruption via ionizing radiation, our device also detects significant amounts of ionizing radiation and triggers the tamper response if detected. To prevent imprinting via long-time storage of the same bit in BBRAM, our software periodically inverts this data. (Carrying out this countermeasure in low-level firmware was tricky; we ended up using free-running counters to produce a random bit.)

An adversary might also compromise security by causing incorrect operation through careful manipulation of various environmental parameters. As a consequence, a device needs to detect and defend against such attacks. One such environmental parameter is supply voltage, which we monitored for several thresholds. For example, at each power-down, the voltage will go from an acceptable level to a low voltage, then to no supply voltage. But the detection and response circuitry needs to be always active — so at some point, it has to switch over to battery operation. (A similar transition occurs at power-up.) Whenever the voltage goes below the acceptable operating level of the CPU and its associated circuitry, these components are all held in a reset state until the voltage reaches the operating point. When the voltage reaches the operating point, the circuitry is allowed to run. If the voltage exceeds the specified upper limit for guaranteed correct operation, it is considered a tamper, and the tamper circuitry is activated.

Another method by which correct operation can be compromised is by manipulating the clock signals that go to the coprocessor. To defend against these sorts of problems, we use *phase locked loops* and independently generated internal clocks to prevent clock signals with missing or extra pulses, or ones that are either too fast or slow. High temperatures can cause improper operation of the device CPU, and even damage it. So, high temperatures cause the device to be held in reset from the operational limit to the storage limit. Detection of temperature above the storage limit is treated as a tamper event.

In the first generation model, operation time of the modular exponentiation engine was correlated to its data, potentially enabling timing attacks (recall Section 3.3.1). To protect against these, we drove the engine with software that manually implemented blinding. In the second generation model, the engine took constant time.

Many attack strategies emerged after our design, but the form factor of an encapsulated multi-chip module enabled a solid design that resisted them. For example, on-board power management resulted in no noticeable signal emerging in power consumption; the encapsulation provides a Faraday cage that appears to resist EMF side-channels. (However, one never knows what news tomorrow's Slashdot will bring.)

6.3   The Source of Secrets

The previous section discussed how we erase device secrets upon tamper. One might deduce that a natural consequence would be that "knowledge of secrets" implies "device is real and untampered". But for this conclusion to hold, we need more premises:

- the secrets were secret when they were first established;

- the device was real and untampered when its secrets were established;

- weakening of cryptography does not compromise the secrets;

- operation of the device has not caused the secrets to be exposed.

This section discusses how we provide the first three properties. Section 6.4 will discuss how we provide the fourth.

6.3.1   Factory Initialization

As one might naturally suspect, an untampered device authenticates itself as such using cryptographic secrets stored in secure memory. The primary secret is the private half of an RSA or DSA key pair. (Chapter 7 elaborates on the use of this private key.) Some symmetric-key secrets are also necessary for some special cases, as Section 6.6.5 will discuss.

The device key pair is generated at device initialization. To minimize risk of exposure, a device generates its own key pair internally, within the tamper-protected environment and using seeds produced from the internal hardware random number generator. The device holds its private key in secure BBRAM, but exports its public key. An external Certificate Authority adds identifying information about the device and its software configuration, signs a certificate for this device, and returns the certificate to the device.

The device-specific symmetric keys are also generated internally at factory initialization. Clearly, the CA must have some reason to believe that the device in question really is an authentic, untampered device. To address this question—and avoid the risks of physical modification in the shipping channel or at the customer site—we initialize the cards in the factory, as the last step of manufacturing. Although factory initialization removes the risks associated with insecure shipping and storage, it does introduce one substantial drawback: the device must remain within the safe storage temperature range during the shipping process. But when considering the point of initialization, a manufacturer faces a tradeoff between ease of distribution and security: we have chosen security.

6.3.2 Field Operations

Regeneration. An initialized device has the ability to regenerate its key pair. Regeneration frees a device from depending forever on one key pair, or key length, or even cryptosystem. Performing regeneration atomically with other actions, such as reloading the crypto code, also proves useful, as Chapter 7 will discuss. For stronger forward integrity, implementations could combine this technique with expiration dates—or even with forward-secure cryptographic techniques.

To regenerate its key pair, a device does the following:

- create a new key pair from internal randomness,

- use the old private key to sign a *transition certificate* for the new public key, including data such as the reason for the change, and

- atomically complete the change, by deleting the old private key and making the new pair and certificate "official."

The current list of transition certificates, combined with the initial device certificate, certifies the current device private key.

Recertification. The CA for devices can also recertify the device, by atomically replacing the old certificate and possibly empty chain of transition certificates with a single new certificate. Clearly, it would be a good idea for the CA to verify that the claimed public key really is the current public key of an untampered device in the appropriate family. This technique can also frees the CA from depending forever on a single key pair, key length, or even cryptosystem.

Revival. Scenarios arise where the tamper detection circuitry in a device has zeroized its secrets, but the device is otherwise untampered. As discussed above, certain environmental changes—such as cold storage or bungled battery removal—trigger tamper response in our design, since otherwise these changes would provide an avenue for undetected tamper. Such scenarios are arguably inevitable in many tamper-response designs,since a device cannot easily wait to see if a tamper attempt is successful before responding.

Satisfying an initial commercial constraint of "save hardware whenever possible" requires a way of reviving such a zeroized but otherwise untampered device. However, such a revival procedure introduces a significant vulnerability: how do we distinguish between zeroized but untampered device, and a tampered device?

How do we perform this authentication? As discussed earlier, we cannot rely on physical evidence to determine whether a given card is untampered, since we fear that a dedicated, well-funded adversary could modify a device (e.g., by changing the internal FLASH components) and then re-assemble it sufficiently well that it passes direct physical inspection. Indeed, the need for factory-initialization was driven by this concern: We can only rely on secrets in tamper-protected secure memory to distinguish a real device from a tampered device.

The problem is basically unsolvable—how can we distinguish an untampered but zeroized card from a tampered reconstruction, when, by definition, every aspect of the untampered card is visible to a dedicated adversary? To accommodate both the commercial and security constraints, our architecture compromises.

- First, we make revival possible. We provide a way for a trusted authority to revive an allegedly untampered but zeroized card, based on authentication via non-volatile, non-zeroizable "secrets" stored inside a particular device component. Clearly, this technique is risky, since a dedicated adversary can obtain a device's revival secrets via destructive analysis of the device, and then build a fake device that can spoof the revival authority.

- We also make revival safe. To accommodate the above risk, we force revival to atomically destroy all secrets within a device, and to leave it without a certified private key. A trusted CA must then re-initialize the device, before the device can "prove" itself genuine to other relying parties. This initialization requires the creation of a new device certificate, which provides the CA with an avenue to explicitly indicate the card has been revived (e.g., "if it produces signatures that verify against Device Public Key N , then it is allegedly a real, untampered device that has undergone revival—so beware"). Thus, we prevent a device that has undergone this risky procedure from impersonating an untampered device that has never been zeroized and revived.

Furthermore, given the difficulty of effectively authenticating an untampered but zeroized card, and the potential risks of a mistake, the support team for the commercial product has decided not to support this option in practice.

6.3.3 Trusting the Manufacturer

A discussion of untamperedness leads to the question: why should the user trust the manufacturer of the device? Considering this question gives rise to three sets of issues.

- **Contents.** Does the black box really contain the advertised circuits and firmware? The paranoid user can verify this probabilistically by physically opening and examining a number of devices. The necessary design criteria and object code listings could be made available to customers under special contract.

- **CA Private Key.** Does the factory CA ever certify bogus devices? Such abuse is a risk with any public key hierarchy. But, the paranoid user can always establish their own key hierarchy, and then design applications that accept as genuine only those devices with a secondary certificate from this alternate authority.

- **Initialization.** Was the device actually initialized in the advertised manner? Given the control a manufacturer might have, it is hard to see how we can conclusively establish that the initialization secrets in a card are indeed relics of the execution of the correct code. However, the cut-and-examine approach above can convince a paranoid user that the key creation and management software in an already initialized device is genuine. This assurance, coupled with the regeneration technique above, provides a solution for the paranoid user: causing their device to regenerate after shipment gives it a new private key that must have been produced in the advertised safe fashion.

6.4 Software Threats

Section 6.2 discussed how we ensure that the core secrets are zeroized upon physical attack, and Section 6.3 discussed how we ensure that they were secret to begin with. However, these techniques still leave an exposure: did the device secrets remain secret throughout operation?

For example, suppose a few months after release, some penetration specialists discover a hole in the OS that allows untrusted user code to execute with full supervisor privilege. Our code loading protocol (Section 6.6) allows us to ship out a patch, and a device installing this patch can sign a receipt with its private key. One might suspect verifying this signature would imply the hole has been patched in that device. Unfortunately, this conclusion would be wrong: a hole that allows untrusted code full privileges would also grant it access to the private key—that is, without additional hardware countermeasures. This section discusses the countermeasures we use.

6.4.1   Software Threat Model

This risk is particularly dire in light of the commercial constraints of multiple layers of complex software, from multiple authorities, remotely installed and updated in hostile environments. History shows that complex systems are, quite often, permeable. Consequently, we address this risk by assuming that all rewritable software in the device may behave arbitrarily badly.

Drawing our defense boundary here frees us from the quagmire of having low-level Miniboot code evaluate incoming code for safety. It also accommodates the wishes of system software designers who want full access to *Ring 0* (that is, "kernel mode"; maximal privileges) in the underlying Intel x86 CPU architecture.

Declaring this assumption often raises objections from systems programmers. We pro-actively raise some counterarguments. First, although all code loaded into the device is somehow "controlled," we need to accommodate the pessimistic view that "signed code" means, at best, good intentions. Second, although an OS typically provides two levels of privilege, history is full of examples where low-level programs usurp higher-level privileges. Finally, as implementers ourselves, we need to acknowledge the very real possibility of error by accommodating mistakes as well as malice.

6.4.2   Hardware Access Locks

In order to limit the abilities of rogue but privileged software, we used *hardware locks:* independent circuitry that restricts the activities of code executing on the main CPU. We chose to use a simple hardware approach for several reasons, including:

■ We cannot rely on the device operating system, since we do not know what it will be—and a corrupt or faulty OS might be what we need to defend against.

■ We cannot rely on the protection rings of the device CPU, because the rewritable OS and Miniboot layers require maximal CPU privilege.

Figure 5.1 (in Section 5.4) shows how the hardware locks fit into the overall design: the locks are independent devices that can interact with the main CPU, but control access to the FLASH and to BBRAM.

However, this approach raises a problem. Critical memory needs protection from bad code. How can our hardware protection—which needs to be simple, if we are to get it right the first time—distinguish between good code and bad code?

We considered and discarded two options:

■ **False Start:** Good code could write a password to the lock.

Although this approach simplifies the necessary circuitry, we had doubts about effectively hiding the passwords from rogue software.

■ **False Start:** The lock determines when good code is executing by monitoring the address bus during instruction fetches.

This approach greatly complicates the circuitry. We felt that correct implementation would be difficult, given the complexities of instruction fetching in modern CPUs, and the subtleties involved in detecting not just the address of an instruction, but the context in which it is executed. For example, it is not sufficient merely to recognize that a sequence of instructions came from the address range for privileged code; the locks would have to further distinguish between several similar scenarios:

- these instructions, executing as privileged code,
- these instructions, executing as a subroutine called by unprivileged code;
- these instructions, executing as privileged code, but with a sabotaged interrupt table.

Solution: Sequence-Based Ratchet. We finally developed a lock approach based on the observation that (because our TCP is a multi-chip module, with physical encapsulation that limits external interaction, like triggering reset, to well-defined and well-regulated channels) *reset* causes all device circuitry return to a known state—and forces the device CPU to begin execution from a fixed address in ROM: known, trusted, permanent code. As execution proceeds, it passes through a non-repeating sequence of code blocks with different levels of trust, permanence, and privilege requirements.

■ Reset starts Miniboot 0, which resides in ROM.

■ Miniboot 0 passes control to Miniboot 1, and never executes again (until the next reboot).

■ Miniboot 1 passes control to the OS, and never executes again.

■ The OS may perform some start-up code.

■ While retaining supervisor control, the OS may then execute application code.

■ The application (executing under control of the OS) may itself do some start-up work, then potentially incur dependence on less trusted code or input.

Our lock design models this sequence with what I called a *trust ratchet,* represented as a nonnegative integer. A small microcontroller stores the the ratchet value in a register. Upon hardware reset, the microcontroller resets the ratchet to 0; through interaction with the device CPU, the microcontroller can advance the ratchet—but can never turn it back. As each block finishes its execution, it advances the ratchet to the next appropriate value. Our implementation also enforces a maximum ratchet value, and ensures that ratchet cannot be advanced beyond this value. models the execution sequence. The microcontroller then grants or refuses memory accesses, depending on the current ratchet value.

Decreasing Trust. The effectiveness of this trust ratchet critically depends on two facts:

- The code blocks can be organized into a hierarchy of decreasing privilege levels (e.g., like classical work in protection rings or lattice models of information flow).

- In our software architecture, these privilege levels strictly decrease in real time after reset.

This time sequencing, coupled with the independence of the lock hardware from the CPU and the fact that the hardware design and its physical encapsulation forces any reset of the locks to also reset the CPU, give the ratchet its power:

- The only way to get the most-privileged level (*Ratchet 0*) is to force a hardware reset of the entire system, and begin executing Miniboot 0 from a hardwired address in ROM, in a known state.

- The only way to get a non-maximal privilege level (*Ratchet* N , for N > 0) is to be passed control by code executing at an earlier, higher-privileged ratchet level.

- Neither rogue software nor any other software can turn the ratchet back to an earlier, higher-privileged level, short of resetting the entire system.

The only avenue for rogue software at Ratchet N to steal the privileges of ratchet K < N would be to somehow alter the software that executes at ratchet K or earlier. However, as Section 6.5 will show, we use the ratchet to prevent these attacks as well.

Generalizations. Although this discussion used a simple total order on ratchet values, nothing prevents using a partial order. Indeed, as we discuss later, our initial implementation of the microcontroller firmware did just that, in order to allow for some avenues for future expansion. (Such expansion never happened, though.)

	Ratchet 0	Ratchet 1	Ratchet 2	Ratchet 3	Ratchet 4
Protected Page 0	Read, Write	No access	No access	No access	No access
Protected Page 1	Read, Write	Read, Write	No access	No access	No access
Protected Page 2	Read, Write	Read, Write	Read, Write	No access	No access
Protected Page 3	Read, Write	Read, Write	Read, Write	Read, Write	No access

Table 6.1. The hardware ratchet locks prevent less trustworthy software entities from reading or writing the protected RAM belonging to more trustworthy software entities.

6.4.3 Privacy and Integrity of Secrets

The hardware locks enable us to address the challenge: how do we keep rogue software from stealing or modifying critical authentication secrets? We do this by establishing *protected pages*: regions of battery-backed RAM which are locked once the ratchet advances beyond a certain level. (The term "page" here refers solely to a particular region of BBRAM—and not to special components of any particular CPU or OS memory architecture.) The hardware locks can then permit or deny write access to each of these pages. Rogue code might still issue a read or write to that address, but the memory device itself will never see it.

Table 6.1 illustrates the access policy we chose: each ratchet level R (for $0 \le R \le 3$) has its own *protected page*, with the property that page P can only be read or written in ratchet level $R \le P$.

We use the term *lockable BBRAM (LBBRAM)* to refer to the portion of BBRAM consisting of the protected pages. (As with all BBRAM in the device, these regions preserve their contents across periods of no power, but zeroize their contents upon tamper.) We ended up using these pages for outbound authentication (Chapter 7); page 0 also holds some secrets used for ROM-based loading (Section 6.6). We partition the remainder of BBRAM into two regions: one belonging to the OS exclusively, and one belonging to the application. Within this nonlockable BBRAM, we expect the OS to protect its own data from the application's.

6.5 Code Integrity

The previous sections presented how our architecture ensures that secrets remain accessible only to allegedly trusted code, executing on an untampered device. To be effective, our architecture must integrate these defenses with techniques to ensure that this executing code really is trusted.

This section presents how we address the problem of code integrity:

- Section 6.5.1 and Section 6.5.2 describe how we defend against code from being formally modified, except through the official code loading procedure.

- Section 6.5.3 and Section 6.5.4 describe how we defend against modifications due to other types of failures.

- Section 6.5.5 summarizes how we knit these techniques together to ensure the device securely boots.

6.5.1   Loading and Cryptography

We confine to Miniboot the tasks of deciding and carrying out alteration of code layers. Although previous work considered a hierarchical approach to loading, our commercial requirements—multiple-layer software, controlled by mutually suspicious authorities, updated in the hostile field, while sometimes preserving state—led to trust scenarios that were simplified by centralizing trust management.

Miniboot 1 (in rewritable FLASH) contains code to support public key cryptography and hashing, and carries out the primary code installation and update tasks—which include updating itself.

Miniboot 0 (in boot-block ROM) contains primitive code to perform DES using the DES-support hardware, and uses secret-key authentication to perform the emergency operations necessary to repair a device whose Miniboot 1 does not function.

6.5.2   Protection against Malice

As experience in vulnerability analysis reveals, practice often deviates from policy. Without additional countermeasures, the policy of "Miniboot is in charge of installing and updating all code layers" does not necessarily imply that "the contents of code layers are always changed in accordance with the design of Miniboot, as installed." For example:

- Without sufficient countermeasures, malicious code might itself rewrite code layers.

- Without sufficient countermeasures, malicious code might rewrite the Miniboot 1 code layer, and cause Miniboot to incorrectly "maintain" other layers.

To ensure that practice meets policy, we use the trust ratchet (Section 6.4) to guard rewriting of the code layers in rewritable FLASH. We group sets of FLASH sectors into *protected segments,* one for each rewritable layer of code. (As with "protected page," the term "segment" is used here solely to denote to these sets of FLASH sectors—and not to special components of a CPU or OS memory architecture.) The hardware locks can then permit or deny write

	Ratchet 0	Ratchet 1	Ratchet 2	Ratchet 3	Ratchet 4
Protected Segment 1	Read, Write, Erase	Read, Write, Erase	Read	Read	Read
Protected Segment 2	Read, Write, Erase	Read, Write, Erase	Read	Read	Read
Protected Segment 3	Read, Write, Erase	Read, Write, Erase	Read	Read	Read

Table 6.2. The hardware ratchet locks prevent less trustworthy software entities from writing (or erasing) the FLASH regions where the code for the more trustworthy software entities live.

access to each of these segments—rogue code might still issue a write to that address, but the memory device itself will never see it.

Table 6.2 illustrates the write policy we chose for protected FLASH. We could have limited Ratchet 0 write-access to Segment 1 alone, since (in practice) Miniboot 0 only writes Miniboot 1. However, it makes little security sense to withhold privileges from earlier, higher-trust ratchet levels—since the earlier-level code could always usurp these privileges by advancing the ratchet without passing control.

As a consequence of applying hardware locks to FLASH, malicious code cannot rewrite code layers unless it modifies Miniboot 1. But this is not possible—in order to modify Miniboot 1, an adversary has to either alter ROM, or already have altered Miniboot 1. Note these safeguards apply only in the realm of attacks that do not result in zeroizing the device. An attacker could bypass all these defenses by opening the device and replacing the FLASH components—but we assume that the defenses of Section 6.2 would ensure that such an attack would trigger tamper detection and response.

In order to permit changing to a hierarchical approach without changing the hardware design, the implemented lock firmware permits Ratchet 1 to advance instead to a Ratchet 2', that acts like Ratchet 2, but permits rewriting of Segment 3. Essentially, our trust ratchet, as implemented, already ranged over a non-total partial order.

6.5.3 Protection against Reburn Failure

In our current hardware implementation, multiple FLASH sectors make up one protected segment. Nevertheless, we erase and rewrite each segment as a whole, in order to simplify data structures and to accommodate future hardware with larger sectors.

This decision leaves us open to a significant risk: a failure or power-down might occur during the non-zero time interval between the time Miniboot starts erasing a code layer to be rewritten, and the time that the rewrite successfully completes. This risk gets even more interesting, in light of the fact that rewrite of a code layer may also involve changes to other state variables and LBBRAM fields.

When crafting the design and implementation, we followed the rule that the system must remain in a safe state no matter what interruptions occur during operations. As we discussed back in Section 3.2.5, this principle is especially relevant to the process of erasing and reburning software resident in FLASH.

- Since Miniboot 1 carries out loading and contains the public key crypto support, we allocate two regions for it in FLASH Segment 1, so that the old copy exists and is usable up until the new copy has been successfully installed. This approach permits using Miniboot 1 for public-key-based recovery from failures during Miniboot 1 updates.

- When reburning the OS or an application, we temporarily demote its state, so that on the next reset after a failed reburn, Miniboot recognizes that the FLASH layer is now unreliable, and cleans up appropriately.

For more complex transitions, we extend this approach: all changes atomically succeed together, or fail either back to the original state, or to a safe intermediate failure state.

6.5.4   Protection against Storage Errors

Hardware locks on FLASH protect the code layers from being rewritten maliciously. However, bits in FLASH devices (even in boot block ROM) can change without being formally rewritten—due to the effects of random hardware errors in these bits themselves.

To protect against spurious errors, we include a 64-bit DES-based MAC with each code layer. Miniboot 0 checks itself before proceeding; Miniboot 0 checks Miniboot 1 before passing control; Miniboot 1 checks the remaining segments. The use of a 64-bit MAC from CBC-DES was chosen purely for engineering reasons: it gave a better chance at detecting errors over datasets the size of the protected segments than a single 32-bit CRC, and was easier to implement even in ROM, given the presence of DES hardware, than more complex CRC schemes.

We reiterate that we do not rely solely on single-DES to protect code integrity. Rather, our use of DES as a checksum is solely to protect against random storage errors in a write-protected FLASH segment. An adversary might exhaustively find other executables that also match the DES MAC of the correct code; but in order to do anything with these executables, the adversary must get write-access to that FLASH segment—in which case, the adversary also has write-access to the checksum, so his exhaustive search was unnecessary.

6.5.5   Secure Bootstrapping

To ensure secure bootstrapping, we use several techniques together:

- The hardware locks on FLASH keep rogue code from altering Miniboot or other code layers.

- The loading protocols (Section 6.6) keep Miniboot from burning adversary code into FLASH.

- The checksums keep the device from executing code that has randomly changed.

If an adversary can cause (e.g., through radiation) extensive, deliberate changes to a FLASH layer so that it still satisfies the checksum it stores, then he can defeat these countermeasures. However, we believe that the physical defenses of Section 6.2 would keep such an attack from being successful:

- The physical shielding in the device would make it nearly impossible to produce such carefully focused radiation.

- Radiation sufficiently strong to alter bits should also trigger tamper response.

Consequently, securely bootstrapping a custom-designed, tamper-protected device is easier than the general problem of securely bootstrapping a general-purpose, exposed machine.

Execution Sequence.   Our boot sequence follows from a common-sense assembly of our basic techniques. Hardware reset forces execution to begin in Miniboot 0 in ROM. Miniboot 0 begins with *Power-on Self Test (POST)* 0, which evaluates the hardware required for the rest of Miniboot 0 to execute. Miniboot 0 verifies the MACs for itself and Miniboot 1. If an external party presents an alleged command for Miniboot 0 (e.g., to repair Miniboot 1), Miniboot 0 will evaluate and respond to the request, then halt. I Layer 1 is not reliable, Miniboot 0 will also halt. Otherwise, Miniboot 0 advances the trust ratchet to 1, and jumps to Miniboot 1.

Except for some minor, non-secret device-driver parameters, no DRAM state is saved across the Miniboot 0 to Miniboot 1 transition. (In either Miniboot, any error or stateful change causes it to halt, in order to simplify analysis. Interrupts are disabled.)

Miniboot 1 begins with POST 1, which evaluates the remainder of the hardware. Miniboot 1 also verifies MACs for Layer 2 and Layer 3. If an external party presents an alleged command for Miniboot 1 (e.g.,to reload Layer 2), Miniboot 1 will evaluate and respond to the request, then halt. Otherwise Miniboot 1 advances the trust ratchet to 2, and if Layer 2 is reliable jumps to Layer 2, the OS.

The OS then proceeds with its bootstrap. If the OS needs to protect data from an application that may find holes in the OS, the OS can advance the trust ratchet to 3 before invoking Layer 3 code. Similarly, the application can advance the ratchet further, if it needs to protect its private data. (Chapter 7 will discuss some OS-level uses we ended up making of the ratchet.)

6.6   Code Loading

One of the last remaining pieces of our architecture is the secure installation and update of trusted code.

In order to accommodate our overall goal of enabling widespread development and deployment of secure coprocessor applications, we needed to consider the practical aspects of this process. We review the principal constraints:

- **Shipped empty.** In order to minimize variations of the hardware and to accommodate US export regulations, it was decided that all devices would leave the factory with only the minimal software configuration (Miniboot only). The manufacturer does not know at ship time (and may perhaps never know later) where a particular device is going, and what OS and application software will be installed on it.

- **Impersonal broadcast.** To simplify the process of distributing code, the code-loading protocol should permit the process to be one-round (from authority to device), be impersonal (the authority does not need to customize the load for each device), and have the ability to be carried out on a public network.

- **Updatable.** As discussed earlier, we needed to be able to update code already installed in devices.

- **Minimal disruption.** An emphatic customer requirement was that, whenever reasonable and desired, application state be preserved across updates.

- **Recoverable.** We needed to be able to recover an untampered device from failures in its rewritable software—which may include malicious or accidental bugs in the code, as well as failures in the FLASH storage of the code or interruption of an update.

- **Loss of cryptography.** The complexity of public key cryptography and hashing code forced it to reside in a rewritable FLASH layer—so the recoverability constraint also implies secure recoverability without these abilities.

- **Mutually suspicious, independent authorities.** In any particular device, the software layers may be controlled by different authorities who may not trust each other, and may have different opinions and strategies for software update.

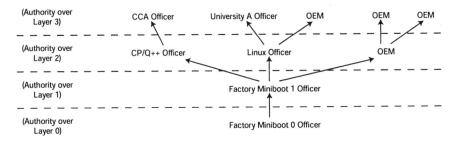

Figure 6.1. We organized the authorities over software layers (over all devices) into a tree; each authority selects and names the authorities who can control the layer above him.

- **Hostile environments.** We can make no assumptions about the user machine itself, or the existence of trusted couriers or trusted security officers.

To address these constraints, we developed and followed some guidelines:

- We make sure that Miniboot keeps its integrity, and that only Miniboot can change the other layers.

- We ensure that the appropriate authorities can obtain and retain control over their layers—despite changes to underlying, higher-trust layers.

- We use public key cryptography whenever possible.

Below, Section 6.6.1 outlines who can be in charge of installing and changing code. Section 6.6.2 discusses how a device can authenticate them. Section 6.6.3 discusses how an "empty" card in the hostile field can learn who is in charge of its code layers. Section 6.6.4 and Section 6.6.5 discuss how the appropriate authorities can authorize code installations and updates.
process with a simple example.

6.6.1   Authorities

As Figure 6.1 illustrates, we organized *software authorities*—parties who might authorize the loading of new software—into a tree. The root is the sole owner of Miniboot; the next generation are the authorities of different operating systems; the next are the authorities over the various applications that run on top of these operating systems. We stress that these parties are external entities, and apply to the entire family of devices, not just one.

Hierarchy in software architecture implies dependence of software. The correctness and security of the application layer depends on the correctness and security of the operating system, which in turn depends on Miniboot 1, which in turn depends on Miniboot 0. (This relation was implied by the decreasing privileges of the trust ratchet.)

Similarly, hierarchy in the authority tree implies dominance: the authority over Miniboot dominates all operating system authorities; the authority over a particular operating system dominates the authorities over all applications for that operating system.

6.6.2   Authenticating the Authorities

Public key authentication. Wherever possible, a device uses a public key signature to authenticate a message allegedly from one of its code authorities. The public key against which this message is verified is stored in the FLASH segment for that code layer, along with the code and other parameters (see Figure 6.2).

Using public key signatures makes it possible to accommodate the "impersonal broadcast" constraint. Storing an authority's public key along with the code in the FLASH layer owned by that authority, enables the authority to change its key pair over time, at its own discretion.

However, effectively verifying such a signature requires two things:

- the code layer is already loaded and still has integrity (so the device actually knows the public key to use); and

- Miniboot 1 still functions (so the device knows . what to do with this public key).

These facts create the need for two styles of loading:

- *ordinary loading*, when these conditions both hold; and

- *emergency loading*, when at least one fails.

Secret-key authentication.. The lack of public key cryptography forces the device to use a secret-key handshake to authenticate communications from the Miniboot 0 authority. The shared secrets are stored in Protected Page 0, in LBBRAM.

Such a scheme requires that the authority share these secrets. Our scheme reconciles this need with the no-databases requirement by having the device itself store a signed, encrypted message from the authority to itself. During factory initialization, the device itself generates the secrets and encrypts this message; the authority signs the message and returns it to the device for safe-keeping. During authentication, the device returns the message to the authority.

6.6.3   Ownership

Clearly, our architecture has to accommodate the fact that each rewritable code layer may have contents that are either reliable or unreliable. However, in order to provided the necessary configuration flexibility, the OS and application

Figure 6.2. A layer contains code, the public key of the authority over that layer, other identifying parameters, and an integrity check.

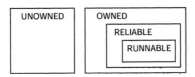

Figure 6.3. The OS and applications layers may be *owned* or *unowned*; an owned layer may also be *reliable*, or even *reliable* and *runnable*.

layers each have additional parameters, reflecting which external authority is in charge of them.

Our architecture addressed this need by giving each of these layers the state space sketched in Figure 6.3:

- The code layer may be *owned* or *unowned*.

- The contents of an owned code layer may be *reliable*: containing useful data. However, some owned layers—and all unowned ones—are *unreliable*.

- A reliable code layer may actually be *runnable*: in a position to execute. However, some reliable layers may be *unrunnable*, for various reasons. Furthermore, all unreliable layers are unrunnable.

This code is stored in EEPROM fields in the hardware lock, write-protected beyond Ratchet 1.

For $0 < N < 3$, the authority over Layer N in a device can issue a Miniboot command giving an unowned Layer $N + 1$ to a particular authority. For $2 \le N \le 3$, the authority over Layer N can issue a command surrendering ownership—but the device can evaluate this command only if Layer N is currently reliable. (Otherwise, the device does not know the necessary public key.)

6.6.4 Ordinary Loading

Code Layer N, for $1 \le N \le 3$, is rewritable. Under ordinary circumstances, the authority over Layer N can update the code in that layer by issuing an update command signed by that authority's private key. This command includes the new code, a new public key for that authority (which could be the same as the old one, per that authority's key policy), and target information to identify the

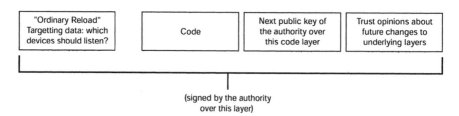

(signed by the authority
over this layer)

Figure 6.4. An ordinary load command for Layer N consists of the new code, new public key, and trust parameters, signed by the authority over that layer; this signature is evaluated against the public key currently stored in that layer.

devices for which this command is valid. The device (using Miniboot 1) then verifies this signature directly against the public key currently stored in that layer. Figure 6.4 sketches this structure.

Target. The target data included with all command signatures allows an authority to ensure that their command applies only in an appropriate trusted environment. An untampered device will accept the signature as valid only if the device is a member of this set. The authority can verify that the load "took" via a signed receipt from Miniboot (see Chapter 7).

For example, suppose an application developer determines that version 2 of a particular OS has a serious security vulnerability. Target data permits this developer to ensure that the untampered devices that will load their application will have version 3 or greater of that operating system.

Underlying Updates. The OS has complete control over the application, and complete access to its secrets; Miniboot has complete control over both the OS and the application. This control creates the potential for serious backdoors. For example, can the OS authority trust that the Miniboot authority will always ship updates that are both secure and compatible? Can the application authority trust that the OS authority uses appropriate safeguards and policy to protect the private key he or she uses to authorize software upgrades?

To address these risks, we permit Authority N to include, when loading its code, *trust parameters* expressing how it feels about future changes to each rewritable layer K < N . In our initial implementation, these parameters only had three values:

- always trust

- never trust, or

- trust only if the update command for K is countersigned by Authority N .

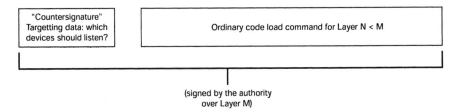

(signed by the authority
over Layer M)

Figure 6.5. An ordinary load command for Layer N can include an optional *countersignature* by the authority over Layer M > N.

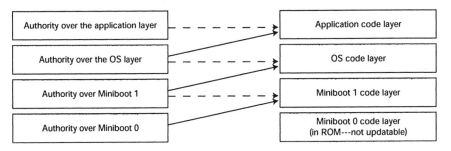

Figure 6.6. Ordinary loading of code into a layer is directly authenticated by the authority over that layer (dashed arrows); emergency loading is directly authenticated by the authority underlying that layer (solid arrows).

(Although we had intended for future expansion, these options turned out to be sufficient—in fact, perhaps more than sufficient; just "always" and "never" probably would have sufficed.)

As a consequence, an ordinary load of Layer N can be accompanied by, for N < M ≤ 3, a countersignature from Authority M, expressing Authority M's opinion about compatibility of that load with some Layer M that he might control. Figure 6.5 sketches this structure.

Update Policy. Trust parameters and countersignatures help us balance the requirements to support hot updates, against the risks of dominant authorities replacing underlying code.

An ordinary reload of Layer N, if successful, preserves the current secrets of Layer N and leaves Layer N runnable.

For N < M ≤ 3, an ordinary reload of Layer N, if successful, preserves the current secrets of Layer M if and only if Layer M had been reliable, and either:

- its trust parameter for N was "always," or

- its trust parameter for N was countersigned, and a valid countersignature from M was included.

Otherwise, the secrets of M are atomically destroyed with the update.

An ordinary load of a layer always preserves that layer's secrets, because presumably an authority can trust their own private key.

6.6.5   Emergency Loading

As observed earlier, evaluating Authority N 's signature on a command to update Layer N requires that Layer N have reliable contents. Many scenarios arise where Layer N will not be reliable—including the initial load of the OS and application in newly shipped cards, and repair of these layers after an interruption during reburn.

Consequently, we require an *emergency* method to load code into a layer without using the contents of that layer. As Figure 6.6 shows, an emergency load command for Layer N must be authenticated by Layer N − 1. As discussed below, our architecture includes countermeasures to eliminate the potential backdoors this indirection introduces.

OS, Application Layers. To emergency load the OS or Application layers, the authority signs a command similar to the ordinary load, but the authority underneath them signs a statement attesting to the public key. Figure 6.7 illustrates this. The device evaluates the signature on this emergency certificate against the public key in the underlying segment, then evaluates the main signature against the public key in the certificate.

This two-step process facilitates software distribution: the emergency authority can sign such a certificate once, when the next-level authority first joins the tree. This process also isolates the code and activities of the next-level authority from the underlying authority.

Risks of Siblings. Burning a segment without using the contents of that segment introduces a problem: keeping an emergency load of one authority's software from overwriting installed software from a sibling authority.

We addressed this risk by giving each authority an *ownerID*, assigned by the N − 1 authority when establishing ownership for N , and stored outside the code layer. The public key certificate later used in the emergency load of N specifies the particular ownerID, which the device checks.

Emergency Reloading of Miniboot. Even though we mirror Miniboot 1, recoverability still required that we have a way of burning it without using it, in order to recover from emergencies when the Miniboot 1 code layer does not function.

Since we must use ROM only and not Miniboot 1, we could not use public key cryptography (management decision). Instead, we use mutual authentication

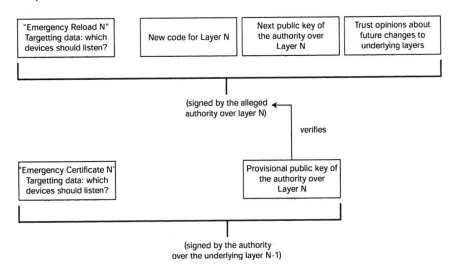

Figure 6.7. An emergency load command for Layer 2 or Layer 3 consists of the new code, new public key for the authority over that layer, and this authority's trust policy for future burns, all signed by the authority over that layer. However, since this is an "emergency," the device cannot assume it knows the public key for that authority. Consequently, the command also includes an emergency certificate signed by the authority over the underlying layer. The main signature is evaluated against the public key in the certificate; the certificate signature is evaluated against the public key stored in the underlying layer.

between the device and the Miniboot 0 authority, based on the device-specific secret keys discussed earlier.

Closing the Backdoors. Emergency loading introduces the potential for backdoors, since reloading Layer N does not require the participation of the authority over that segment. For example, an OS authority could, by malice or error, put anyone's public key in the emergency certificate for a particular application authority.

Since the device cannot really be sure that an emergency load for Layer N really came from the genuine Authority N , Miniboot enforces two precautions:

- It erases the current Layer N secrets but leaves the segment runnable from this clean start (since the alleged owner trusts it).

- It erases all secrets belonging to later layers, and leaves them unrunnable (since their owners cannot directly express trust of this new load).

These actions take place atomically, as part of a successful emergency load.

6.7   Putting it All Together

This architecture establishes individual commands for Authority N to:

- establish owner of Layer N + 1

- attest to the public key of that owner

- install and update code in Layer N ,

- express opinions about the trustworthiness of future changes to Layer K , for K < N .

Except for emergency repairs to Miniboot 1, all these commands are authenticated via public key signatures, can occur over a public network, and can be restricted to particular devices in particular configurations.

We retain a layer's secrets in BBRAM/LBBRAM only while we can continuously maintain an environment for that layer that its owner trusts. We must be able to verify this trust a priori, via public key cryptography.

We illustrate how this architecture supports flexible code development with a simple example. Suppose Alice is in charge of Miniboot 1, and Bob wants to become a Layer 2 owner, in order to develop and release Layer 2 software on some cards. Bob generates his key pair, and gives a copy of his public key to Alice. Alice then does three things for Bob:

- She assigns Bob a 2-byte ownerID value that distinguishes him among all the other children of Alice

- She signs an "Establish Owner 2" command for Bob.

- She signs an "Emergency Signature" for an "Emergency Burn 2" saying that Owner 2 Bob has that public key.

Bob then goes away, writes his code, prepares the remainder of his "Emergency Burn 2" command, and attaches the signature from Alice.

Now, suppose customer Carol wants to load Bob's program into Layer 2 on her card. She first buys a virgin device which has an unowned Layer 2, but has Miniboot 1 and Alice's public key in Layer 1 . Carol gets from Bob his "Establish Owner 2" and "Emergency Burn 2" command, and plays them into her virgin card via Miniboot 1. It verifies Alice's signatures and accepts them. Layer 2 in Carol's card is now owned by Bob, and contains Bob's Program and Bob's Public key.

If Bob wants to update his code and/or key pair, he simply prepares an "Ordinary Burn 2" command, and transmits it to Carol's card. Carol's card checks his signature on the update against the public key it has already stored for him.

Note that Bob never exposes to Alice his private key, his code, his pattern of updates, or the identity of his customers. Furthermore, if Bonnie is another Layer 2 developer, she shares no secrets with Bob, and updates for Bonnie's software will not be accepted by cards owned by Bob's.

The architecture also support other variations in the installation/development process; for example, maybe Bob buys the cards himself, configures them, then ships them to Carol.

The case for Layer 3 developers is similar.

6.8   What's Next

Chapter 7 will discuss how we used this architecture to enable an application running on an untampered TCP to prove it is the "real thing, doing the right thing."

Chapter 8 will expand on the security properties it provided, and how we formalized and evaluated them.

6.9   Further Reading

This chapter was based in part on my 1999 *Computer Networks* paper [SW99]: Section 6.1 through Section 6.7 on Section 3 through Section 9, respectively. As noted earlier, the physical security architecture of Section 6.2 was primarily the work of Steve Weingart.

Chapter 7

OUTBOUND AUTHENTICATION

Physical security, maintaining secrets, and installing code do not by themselves enable solutions to the problems we laid out; the code safely residing in this tamper-protected environment needs to be able to prove who it is to remote parties. This chapter will discuss this *outbound authentication* issue, and the theoretical framework I developed to reason about the problem of how to enable coprocessor applications to participate as full-fledged entities in distributed cryptographic protocols.

This work represented evolution in thinking about TCPs. What's important was not secrecy of code going into the TCP, but the ability of a relying party to tell what's there. However, as the design work proceeded, a further refinement is emerging: what's important is not so much the names of all the elements in the code configuration, but rather whether the relying party should trust it.

Section 7.1 introduces the problem. Section 7.2 presents the theoretical underpinnings, and Section 7.3 discusses the implementation.

7.1 Problem

Using TCPs to secure distributed computation requires *outbound authentication (OA):* the ability of coprocessor applications to authenticate themselves to remote parties. Code downloading loses much of its effect if one cannot easily authenticate the entity that results! (Gasser et al provide an early consideration of this problem in the settings of a general distributed system architecture [GGKL89].)

Merely configuring the coprocessor platform as the appropriate entity (e.g., a rights box, a wallet, an auction marketplace) does not suffice in general. A signed statement *about* the configuration also does not suffice. For maximal effectiveness, the platform should enable the *entity itself* to have authenticated key pairs and to engage in protocols with any party on the Internet: so that only that particular trusted auction marketplace, following the trusted rules, is able to receive the encrypted strategy from a remote client; so that only that particular trusted rights box, following the trusted rules, is able to receive the object and the rights policy it should enforce.

In theory, solutions where the entity does not possess its own key pair but makes use of some other service are also possible. We did not consider them for several reasons. First, having one's own key pair is the universal building block for the main body of distributed security protocols—relying parties draw conclusions based on whether one proves knowledge of a private key. By providing each entity its own key pair, we reduce to barriers to using such protocols with TCP-resident entities. Furthermore, having many entities share a key would overly complicate the API for key usage, and make it harder to have confidence that the design and implementation did not hide bugs.

7.1.1   The Basic Problem

A relying party needs to conclude that a particular key pair really belongs to a particular software entity within a particular untampered platform. As Chapter 5 and Chapter 6 discussed, design and production constraints led to a nontrivial set of software entities in a coprocessor at any one time, and in any one coprocessor over time. For our TCP—and for computation in general— relying parties tend to trust some of these entities and not others. Software and hardware structure can introduce further dependencies that might affect the conclusions relying parties would reach if they knew these dependencies. For example, if a TCP permitted multiple concurrent applications, some of these were hostile, and the OS (through oversight) permitted hostile applications to subvert other ones, then a relying party might want to know additional details about the software configuration surrounding a particular application instance. ("Might someone I do not trust have been in the box at the same time?")

Furthermore, we needed to accommodate a multiplicity of trust sets (different parties have different views), as well as the dynamic nature of any one party's trust set over time. This background sets the stage for the basic problem: how should the device generate, certify, change, store, and delete private keys, so that relying parties can draw those conclusions, and only those conclusions, that are consistent with their trust set?

7.1.2   Authentication Approach

As Chapter 5 discussed, another business constraint we had was that the only guaranteed contact we (as the manufacturer) would have with a TCP was at manufacture time. In particular, we could assume no audits or database of TCP-specific data (secret or otherwise), nor provide any online services to cards once they left the factory. This constraint naturally suggested the use of public key cryptography for authentication, both *inbound* (from the outside world into the TCP) and *outbound* (from inside the TCP back into the outside world). This choice separates TCPs from relying parties and frees the manufacturer from having to track any association of particular platforms with their ultimate location, users, and applications.

Chapter 6 discussed how we handled inbound authentication: pre-installing a public key (in FLASH) so the TCP knows how to verify the first command it receives, and building up from there.

For outbound authentication, the natural approach is to keep a private key in tamper-protected memory and have something create signed certificates about the corresponding public key. Because of the last-touch-at-manufacturing constraint (and because of a design assumption that the manufacturer would be the central trust root for these devices), the last time we can ensure that an external trust point can interact with the TCP and sign such certificates is at the factory. After that, it is up to the TCP itself.

7.1.3 User and Developer Scenarios

Discussions about potential relying parties led to additional requirements.

Developers were not necessarily going to trust each other. For example, although an application developer must trust the contents of the lower layers when his application is actually installed, he should be free to require that his secrets be destroyed should a lower layer be updated in a way he does not trust. These discussions resulted in the update policy features presented in Chapter 6. Each code load can specifying the conditions under which that layer's secrets should be preserved across changes to lower layers. Any other scenario destroys secrets.

However, even those relying parties who wanted the device to preserve secrets across updates reserved the right to change their opinions about whether a particular version of code was trustworthy. A relying party might have trusted some version of code, but post facto decide instead not to trust it—even if it was his own code. Relying parties wanted to be able to verify whether an untrusted version had been installed during the lifetime of secrets they cared about.

In theory, the OS layer should resist penetration by a malicious application; in practice, operating systems have a bad history here, so we only allow one application above it and intend the OS layer solely to assist the application developer. (That is, the OS can support multiple concurrent processes, but we assume these are all in the same trust domain.) Furthermore, we need to allow that some relying parties will believe that the OS in general (or a specific version) may indeed be penetrable by malicious applications.

Small-scale developers (without a large pre-established reputation) may be unable to assure the public of the integrity and correctness of their applications (e.g., through code inspection, formal modeling, etc). Where possible, we should maximize the credibility our architecture can endow on applications from such developers.

We note that this design assumption of "one application space" was driven by the generally poor record of operating systems in this regard and by the lack of a suitable high-assurance candidate at the time of product development. Considering our problem in the framework of a higher assurance operating system, where this restriction may be unnecessary, or a general-purpose desktop, where this restriction may be unacceptable, is an interesting area of future work.

7.1.4   On-Platform Entities

We want to give key pairs to entities that consist of software running an TCP. One of the first things we need to deal with is the notion of what such an entity is. Let's start with a simple case: suppose the TCP had exactly one place to hold software and that the TCP zeroized all state with each code load. In this scenario, the notion of entity is pretty clear: a particular code load C_1 executing inside an untampered device D_1. The same code C_1 inside another device D_2 would constitute a different entity, as would a re-installation of C_1 inside D_1.

However, this simple case raises challenges. If a reload replaces C_1 with C_2, and reloads clear all tamper-protected memory, how does the resulting entity (C_2 on D_1) authenticate itself to a party on the other side of the net? The card itself would have no secrets left since the only data storage hidden from physical attack was cleared. Consequently, any authentication secrets would have to come with C_2, and we would start down a path of shared secrets and personalized code loads.

This line of thinking leads to questions. Should an application entity "include" the OS underneath it? Should it include the configuration control layers that ran earlier in this boot sequence but are no longer around? (As we discuss later, one can even make a case that an entity should include entities that were previously installed but are no longer present on the card.)

Since we built the 4758 to support real applications, we gravitated toward a practical definition: an entity is an installation of the application software in a trusted place, identified by all underlying software and hardware.

7.1.5   Secret Retention

As noted, developers demanded that we sometimes permit secret retention across reload. With a secret-preserving load, the entity may stay the same, but the code may change. The conflicting concepts that developers had about what exactly happens to their on-card entity when a code update occurs led us to think more closely about entity lifetimes. We introduce some language—*epoch* and *configuration*–to formalize that.

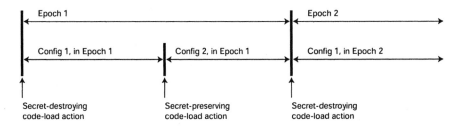

Figure 7.1. An *epoch* starts with a a code-load action that clears a layer's secrets; each code-load that changes that layer—or the layers it uses—but preserves its secrets starts a new *configuration*.

The idea is that the a "configuration" is the exact software stack supporting an entity; every time any of those components changed (or, for that matter, undergoes a "null change" by being reburned with the same contents), we start a new configuration. However, an "epoch" is the length of time that the secrets are preserved.

A Layer N epoch entity thus consists of a sequence of Layer N configuration entities. This sequence may be unbounded—since any particular epoch might persist indefinitely, across arbitrarily many configuration changes, if the code-loading officer included policies that permitted such persistence across such changes.

For example, Layer N may undergo a transition such as a secret-preserving update, a complete reinstall, or an ownership surrender. A hot update will begin a new Layer N configuration but will preserve the old Layer N epoch; whether it preserves a K epoch (for K > N) depends on the policy the owner of Layer K has established.

Figure 7.1 sketches these concepts.

Definition (Configuration, Epoch). *A Layer N configuration is the maximal period in which that Layer is runnable, with an unchanging software environment in Layers 1..N. A Layer N epoch is the maximal period in which the Layer can run and accumulate state. If E is an on-platform entity in Layer N ,*

- E *is an* epoch-entity *if its lifetime extends for a Layer N epoch.*

- E *is a* configuration-entity *if its lifetime extends for a Layer N configuration.*

7.1.6 Authentication Scenarios

This design left us with on-card software entities made up of several components with differing owners, lifetimes, and state. A natural way to do outbound authentication is to give the card a certified key pair whose private key lives in tamper-protected memory. However, the complexity of the entity structure creates numerous problems.

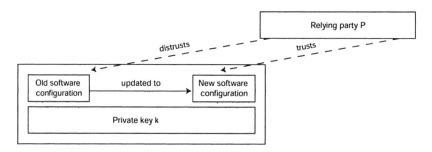

Figure 7.2. Replacing untrusted software with trusted software, while retaining the private key, creates problems. What should the relying party conclude about an entity that proves knowledge of this private key?

Application Code. Suppose entity C is the code C_1 residing in the application Layer 3 in a particular device. C may change. Two possible changes include:

- a simple code update taking the current code C_1 to C_2, or

- a complete reinstall of a different application from a different owner, taking C_1 to C_3.

If a relying party P trusts C_1, C_2, and C_3 to be free of flaws, vulnerabilities, and malice, then the natural approach might work. However, if relying party P distrusts some of this code, then problems arise.

- If relying party P does not trust C_1, then how can P distinguish between an entity with the C_2 patch and an entity with a corrupt C_1 pretending to have the C_2 patch? (See Figure 7.2.)

- If relying party P does not trust C_2, then then how can P distinguish between an entity with the honest C_1 and an entity with the corrupt C_2 pretending to be the honest C_1? (The mere existence of a signed update command compromises all the cards—since the relying party cannot know whether any particular card carried out this update. See Figure 7.3.)

- If relying party P does not trust C_3, then how can P distinguish between the honest C_1 and a malicious C_3 that pretends to be C_1? (Essentially, this is isomorphic to Figure 7.3.)

Code-loading Code. Even more serious problems arise if a corrupted version of the configuration software in Layer 1 exists. If an evil version existed that allowed arbitrary behavior, then (without further countermeasures) a relying party P cannot distinguish between any on-platform entity E_1 and an E_2 consisting of a rogue Layer 1 carrying out some elaborate impersonation.

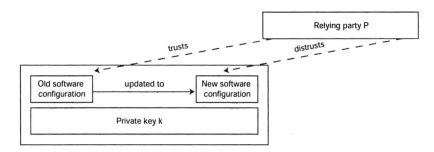

Figure 7.3. The potential to replace trusted software with untrusted software, while retaining the private key, also creates problems. If the relying party P does not know whether or not the update happened, what can P conclude about an entity that proves knowledge of this private key?

OS Code. Problems can also arise because the OS code changes. Debugging an application requires an operating system with debug hooks; in final development stages, a reasonable scenario is to be able to "update" back-and-forth between a version of the OS with debug hooks and a version without. (Indeed, the development toolkit that Section 5.4 discussed does just this—under a special "development" Layer 2 ownerID.)

With no additional countermeasures, a relying party P cannot distinguish between the application running securely with the real OS, the application with debug hooks underneath it, and the application with the real OS but with a policy that permits hot updates to the debug version. The private key would be the same in all cases.

7.1.7 Internal Certification

The above scenarios suggest that perhaps a single key pair, for all entities in a TCP for the lifetime of the TCP, may not suffice. If two different entities, one trusted and one untrusted, had access to the same private key material, then the relying party can no longer draw a reasonable conclusion from use of the private key alone. If we want to enable the relying party to do this, the natural generalization is to have separate keys for separate entities. However, extending to schemes where one on-platform entity generates and certifies key pairs for other on-platform entities also creates challenges.

For example, suppose Layer 1 generates and certifies key pairs for the Layer 2 entity. If a reload replaces corrupt OS B_1 with an honest B_2, then the relying party P should be able to distinguish between the certified key pair for B_2 and that for B_1. However, without further countermeasures, if supervisor-level code can see all data on the TCP, then B_1 can forge messages from B_2—since it could have seen the Layer 1 private key.

A similar penetrated-barrier issue arises if we expect an OS in Layer 2 to maintain a private key separate from an application Layer 3, or if we entertained schemes where mutually suspicious applications executed concurrently. If a hostile application might in theory penetrate the OS protections, then an external party cannot distinguish between messages from the OS, messages from the honest application, and messages from rogue applications.

This line of thinking led us to the more general observation that, if the certifier outlives the certified, then the integrity of what the certified does with their key pair depends on the future behavior of the certifier.

In the case of our coprocessor, this observation has subtle and dangerous implications; for example, one of the reasons we centralized configuration control in Layer 1 was to enable the application developer to distrust the OS developer and request that the application (and its secrets) be destroyed, if the underlying OS undergoes an update the application developer does not trust. What if the untrusted OS has access to a private key used in certifying the original application? (This observation might also have implications for other types of PKI, where a CA/RA both generates as well as certifies user key pairs.)

7.2 Theory

As we have discussed, using certified key pairs seems the natural choice for outbound authentication. However, as we just sketched, the straightforward approach of just sending the TCP out with a certified key pair permits trouble.

In this section, we try to formalize the principles that emerged while considering this problem.

A TCP leaves the factory and undergoes some sequence of code loads and other configuration changes. A relying party interacts with an entity allegedly running inside this TCP. The platform's OA scheme enables this application to wield a private key and to offer a collection of certificates purporting to authenticate its keyholder.

It would be simplest if the party could use a straightforward validation algorithm on this collection. As Maurer [KM00, Mau96] formalized, a relying party's validation algorithm needs to consider which entities that party trusts. Our experience showed that parties have a wide variety of trust views that change dynamically. Furthermore, we saw the existence of two spaces:

- the conclusions that a party *will* draw, given an entity's collection of certificates and the party's trust view, and

- the conclusions that a party *should* draw, given the history of those keyholders and the party's trust view.

We needed to design a scheme that permits these sets of conclusions to match, for parties with a wide variety of trust views.

7.2.1 What the Entity Says

The relying party P wants to authenticate interaction with a particular entity E. For this interaction to be meaningful, P must already trust E to behave correctly with its keys (we will elaborate on this point later). Many scenarios could exist here; for simplicity, our analysis reduces these to the scenario of E needing to prove to P that own (E ,K): that E has exclusive use of the private element of key pair K ; that (in P 's view) no one who might subvert this will do so.

We need to be able to talk about what happens to a particular platform: both a long-term sequence of actions, as well as specific instants along that sequence. So we introduce some notation—*history* and *run*—for these concepts.

A platform can take action only in the context of the particular history H that it has experienced to that point in time. However, we need to consider both history and run because this run may continue in several different ways beyond that point and the actions in these potential futures may be relevant to the conclusions a relying party draws from an action the platform takes now.

Definition (History, Run,). *Let a* history *be a finite sequence of computation for a particular device. Let a* run *be some unbounded sequence of computation for a particular device. We write* H R *when history* H *is a prefix of run* R .

In the context of OA for platforms that cannot be opened or otherwise examined, and that disappear once they leave the factory, it seemed reasonable to impose the restriction that on-platform entities carry their certificates with them. For simplicity, we also imposed the restriction that they present the same fixed set no matter who asks.

Definition. *When entity* E *wishes to prove it owns* K *after history* H , *let* Chain (E ,K ,H) *denote the set of certificates that it presents.*

7.2.2 What the Relying Party Concludes

Will a relying party P believe that entity E owns key pair K ?

First, we need some notion of trust. A relying party P usually has some ideas of which on-platform applications it might trust to behave "correctly" regarding keys and signed statements, and of which ones it is unsure.

Definition. *For a relying party* P , *let* TrustSet (P) *denote the set of entities whose statements about certificates* P *trusts. Let* root *be the factory CA: the trust root for this family of platforms. A* legitimate *trust set is one that contains* root.

As discussed earlier, this project arose in the context of a specific commercial product effort, which imposed some specific constraints. In particular: our design had to assume that the manufacturer could not construct a database of these platforms, nor track could not track where they went. Once the platforms were deployed, we could neither contact nor audit them, nor could we assume that they or their applications r relying parties would have network access back to the factory. These constraints made revocation infeasible. Consequently, for the problem space we faced, it was reasonable to impose the restriction that the external party decides validity based on an entity's chain and the party's own list of trusted entities.

We formalize this notion of "reasonable" validation schemes.

Definition (Trust-set scheme). *A* trust-set *certification scheme is one where the relying party's* Validate *algorithm is deterministic on the variables* Chain (E, K, H) *and* TrustSet (P).

We thus needed to design a trust-set certification scheme that accommodates *any legitimate trust set*, since discussion with developers (and experiences doing security consulting) suggested that relying parties would have a wide divergence of opinions about which versions of which software they trust.

7.2.3   Dependency

The problem scenarios in Section 7.1.6 arose because one entity E_2 had an unexpected avenue to use the private key that belonged to another entity E_1. We need language to express these situations, where the integrity of E_1's key actions *depends* on the correct behavior of E_2.

We formalize this concept as a *dependency function*, taking an entity to the set of entities that can subvert its correct operation, with respect to private keys.

Definition (Dependency Function). *Let* E *be the set of entities. A* dependency function *is a function* $D : E \dashrightarrow 2^E$ *such that, for all* E_1, E_2, *we have:*

- $E_1 \in D(E_1)$ *(Idempotency)*

- *if* $E_2 \in D(E_1)$ *then* $D(E_2) \subset D(E_1)$ *(Transitivity)*

When a dependency function depends on the run R, *we write* D_R.

Different entity architectures give rise to different appropriate dependency functions.

In our specialized hardware, code runs in a single-sandbox controlled environment which (if the physical security works as intended) is free from outside observation or interference. Hence, in our analysis, dependence should follow from the ability of an entity to read or write another entity's secrets, or to modify code that can read or write another entity's secrets.

Definition. *For entities* E_1 *and* E_2 *in run* R:

- *We write* $E_2 \xrightarrow{\text{data}}_R E_1$ *when* E_1 *has read/write access to the secrets of* E_2. ($E_2 \xrightarrow{\text{data}}_R E_2$ *trivially.*)

- *We write*

$$E_2 \xrightarrow{\text{code}}_R E_1$$

  when E_1 *has write access to the code of* E_2.

- *Let* $\rightarrow_R$ *be the transitive closure of the union of these two relations.*

- *For an entity* E *in a run* R, *define*

$$\text{Dep}_R(E) = \{F : E \rightarrow_R F\}$$

The intuition here is that, for the platform architecture we considered, $Dep_R(E)$ lists all the on-platform software entities that could have subverted the correct operation of entity E in run R.

In terms of our coprocessor, if C_1 follows B_1 in the post-boot sequence, then we have $C_1 \xrightarrow{data}_R B_1$ (since B_1 could have manipulated data before passing control). If C_2 is a secret-preserving replacement of C_1, then $C_1 \xrightarrow{data}_R C_2$ (because C_2 still can touch the secrets C_1 left). If A can reburn the FLASH segment where B lives, then $B \xrightarrow{code}_R A$ (because A can insert malicious code into B, that would have access to B's private keys).

7.2.4   Soundness

Should the relying party draw the conclusions it actually will? In our analysis, security dependence depends on the run; entity and trust do not. This leads to a potential conundrum. Suppose, in run R, we have:

- $C \rightarrow_R B$ and

- $C \in \text{TrustSet}(P)$, but

- $B \notin \text{TrustSet}(P)$.

Then a relying party P cannot reasonably accept any signed statement from C, because B may have forged it.

To capture this notion, we define *soundness* for OA. The intention of soundness is that if a relying party concludes that a message came from an entity, then it really did come from that entity—modulo the relying party's trust view. The party will not conclude that it *should* trust the entity, if such a conclusion would be inconsistent with the party's beliefs.

That is, suppose in some history $H \quad R$, P concludes own(E, K) from Chain(E, K, H). If the TrustSet(P) entities behave themselves, then E should really own K. We formalize this notion:

Definition. *An OA scheme is sound for a dependency function* D *when, for any entity* E, *a relying party* P *with any legitimate trust set, and any history and* *run* H   R *:*

$$\text{Validate}(P, \text{Chain}(E, K, H)) \implies D_R(E) \subseteq \text{TrustSet}(P)$$

We restrict our attention to legitimate trust sets because given commercial product constraints (a party could not open and examine a platform without destroying it), it would be difficult for a relying party who did not trust root to draw any useful conclusions.

7.2.5   Completeness

One might also ask if the relying party *will* draw the conclusions it actually *should*. We consider this question with the term *completeness*. If in any run where E produces some Chain (E ,K ,H) and $D_R(E)$ is trusted by the relying party P —so in P 's view, no one who had a chance to subvert E would have— then P should conclude that E owns K .

Definition. *An OA scheme is* complete *for a dependency function* D *when, for any entity* E *claiming key* K *, relying party* P *with any legitimate trust set, and history and run* H   R *:*

$$D_R(E) \subseteq \text{TrustSet}(P) \implies \text{Validate}(P, \text{Chain}(E, K, H))$$

Note that by our definition of TrustSet, if $D_R(E) \subseteq \text{TrustSet}(P)$, then P believes that E will act honestly.

7.2.6   Achieving Both Soundness and Completeness

These definitions equip us to formalize a fundamental observation. If we're going build a trust-set authentication scheme that is both sound and complete, then the certificate chain for an entity needs to name its full dependency set. Figure 7.4 sketches why.

Theorem. *Suppose a trust-set OA scheme is both sound and complete for a given dependency function* D *. Suppose entity* E *claims* K *in histories* H_1 R_1 *and* H_2 R_2. *Then:*

$$D_{R_1}(E) = D_{R_2}(E) \implies$$

$$\text{Chain}(E, K, H_1) = \text{Chain}(E, K, H_2)$$

Proof. Suppose $D_{R_1}(E) = D_{R_2}(E)$ but Chain (E ,K ,H_1) = Chain (E ,K ,H_2). We cannot have both $D_{R_1}(E) \subseteq D_{R_2}(E)$ and $D_{R_2}(E) \subseteq D_{R_1}(E)$, so, without loss of generality, let us assume $D_{R_2}(E) \subseteq D_{R_1}(E)$. There thus exists a set S with $D_{R_1}(E) \subseteq S$ but $D_{R_2}(E) \subseteq S$.

Since the scheme is sound and complete, it must work for any legitimate trust set, including S. Let relying party P have S = TrustSet (P). Since this is a trust-set certification scheme and E produces the same chains in both histories,

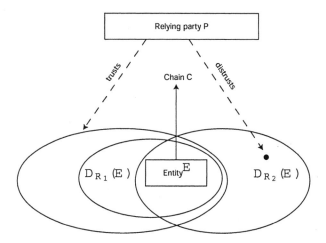

Figure 7.4. If E produces the same chain but may or may not depend on things that P does not trust, then P must accept a chain it should reject, or reject a chain it should accept.

party P must either validate these chains in both scenarios, or reject them in both scenarios. If party P accepts in run R_2, then the scheme cannot be sound for D, since E depends on an entity that P did not trust. But if party P rejects in run R_1, then the scheme cannot be complete for D, since party P trusts all entities on which E depends.   □

7.2.7   Design Implications

We consider the implications of the above theorem for specific ways of constructing chains and drawing conclusions, for specific notions of dependency.

For example, we can express the standard approach—the relying party P makes its conclusion by recursively verifying signatures and applying a basic inference rule—in a Maurer-style calculus [KM00]. Suppose C is a set of *certificates*: statements of the form K_1 says own (E_2, K_2). Let S be the set of entities that P *trusts to speak the truth about assertions of key ownership*. That is:

$$S \quad = \quad \text{TrustSet} (P)$$

We can then start reasoning about $\text{View}_{\text{will}} (C, S)$: the set of certificates statements and key ownership conclusions that this party will conclude are true, given the set of entities the party trusts.

A relying party may start by believing

$$C \cup \{\text{own} (\text{root}, K_{\text{root}})\}$$

So, we initialize $\mathsf{View_{will}}$ (C,S) to that set of statements. We then keep adding statements derivable from this set by applying the rule

$$\mathsf{own}\,(E_1,K_1),\ E_1 \in S,\ K_1\ \mathsf{says}\ \mathsf{own}\,(E_2,K_2)$$

$$\mathsf{own}\,(E_2,K_2)$$

Informally, if the party trusts an entity and believes that entity owns a key, then it believes certificates that key signs. The Validate algorithm for party P then reduces to the decision of whether $\mathsf{own}\,(E,K)$ is in this set.

We can also express what a party should conclude about an entity, in terms of the chain the entity presents, and the views that the party has regarding trust and dependency. If D is a dependency function, we can define $\mathsf{View_{should}}$ (C,S,D) to be the set of statements derivable by applying the alternate rule:

$$\mathsf{own}\,(E_1,K_1),\ D\,(E_1) \subseteq S,\ K_1\ \mathsf{says}\ \mathsf{own}\,(E_2,K_2)$$

$$\mathsf{own}\,(E_2,K_2)$$

Informally, if the party trusts an entity, and the entity (in this run) is in an configuration environment that the party trusts, then the party should believe certificates signed with that entity's key.

In terms of this calculus, we obtain soundness by ensuring that for any chain and legitimate trust set, and $H \quad R$, the set $\mathsf{View_{will}}$ $(\mathsf{Chain}\,(E,K,H),S)$ is contained in the set $\mathsf{View_{should}}$ $(\mathsf{Chain}\,(E,K,H),S,D_R)$. The relying party should only use a certificate to reach a conclusion when the entire dependency set of the signer is in $\mathsf{TrustSet}\,(P)$.

By construction of the inference rules, we can see that containment holds the other way.

$$\mathsf{View_{should}}\,(\mathsf{Chain}\,(E,K,H),S,D_R) \subseteq$$

$$\mathsf{View_{will}}\,(\mathsf{Chain}\,(E,K,H),S)$$

7.3   Design and Implementation

For simplicity of verification, we would like $\mathsf{Chain}\,(E,K,H)$ to be a literal chain: a linear sequence of certificates going back to root. To ensure soundness and completeness, we need to make sure that, at each step in the chain, we maintain the invariant that the partial set of certifiers equals the dependency set of that node (for the dependency function we see relying parties using). To achieve this goal, the elements we can manipulate include generation of this chain, as well as how dependency is established in the device. In particular, we follow two guidelines:

- use the software and hardware architecture to *eliminate* any unnecessary dependence

- and then ensure that the dependency set that remains *participates* in certification

7.3.1 Layer Separation

Because of the post-boot execution sequence, code that executes earlier can subvert code that executes later. (With only one chance to get the hardware right, we did not feel comfortable with attempting to restore the system to a more trusted state, short of reboot.) If B ,C are Layer i,Layer i+ 1 respectively, then C $\dashrightarrow_R$ B unavoidably.

However, the other direction should be avoidable, and (as Section 6.4.3 discussed) and we used hardware ratchet locks to avoid it. To ensure B $\overset{data}{\dashrightarrow}_R$ C , we reserved a portion of BBRAM for B , and used the ratchet hardware to enforce access control (Section 6.4.3). (Essentially, this technique refines and extends—and implements—the key-hiding technique suggested by Lampson et al [LABW92, p. 294].) To ensure B $\overset{code}{\dashrightarrow}_R$ C , we write-protect the FLASH region where B is stored (Section 6.5.2). The ratchet hardware restricts write privileges only to the designated prefix of this execution sequence.

To keep configuration entities from needlessly depending on the epoch entities, in our Model 2 device, we subdivided the higher BBRAM to get four regions, one each for epoch and configuration lifetimes, for Layer 2 and Layer 3. The initial boot-time clean-up code Layer 1 (already in the dependency set) zeroizes the appropriate regions on the appropriate transition. That is, if this boot sequence does not preserve the Layer K epoch, the BBRAM region for the Layer K epoch is zeroized; if this boot sequence does not preserve the Layer K configuration, the BBRAM region for the Layer K configuration is zeroized.

(For transitions to a new Layer 1, the clean-up is enforced by the *old* Layer 1 and the permanent Layer 0—to avoid incurring dependency on the new code.)

7.3.2 The Code-Loading Code

As discussed elsewhere, we felt that centralizing code-loading and policy decisions in one place enabled cleaner solutions to the trust issues arising when different parties control different layers of code. But this centralization creates some issues for OA. Suppose the code-loading Layer 1 entity A_1 is reloaded with A_2. As Section 6.6.4 discussed, constraints dictated that A_1 itself do the reloading, because the ROM code below it had no public key support. It is unavoidable that $A_2 \overset{code}{\dashrightarrow}_R A_1$ (because A_1 could have cheated, and not installed the correct code). However, to avoid $A_1 \overset{data}{\dashrightarrow}_R A_2$, we take these steps as an atomic part of the reload: A_1 generates a key pair for its successor A_2; A_1 uses its current key pair to sign a *transition certificate* attesting to this change of versions and key pairs; and A_1 destroys its current private key. Figure 7.5 illustrates this process.

This technique—which we implemented and shipped with the Model 1 devices in 1997—differs from the concept of *forward security* [Andb, Gun90] in

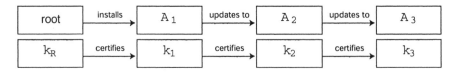

Figure 7.5. When the code-loading layer updates itself, it generates and certifies a new key pair for its successor.

that we change keys with each new version of software, and ensure that *the name of the new version is spoken by the old version.* That is: the device leaves the factory with a key pair owned by Layer 1, and a certificate signed by the factory root that names the factory root, and that binds that public key to that device (specified by model number, serial number, etc) with that version of Layer 1 code (identity of the owner of this layer, name they gave to this code, revision number they gave to this code, SHA-1 hash of this code, when the current Layer 1 epoch started, when the current Layer 1 configuration started, etc.).   Each such update of Layer 1, then adds a transition certificate, signed by the old version, which names both the old and new version of the Layer 1 code, as well as the fact that a transition took place from the old to the new.

(The BirliX security architecture proposed having an on-platform entity generate and certify a key pair with each boot [HKK93, Section 6]; this concept also fits into our framework.)

As a consequence, a single malicious version cannot hide its presence in the trust chain; for a coalition of malicious versions (and the rest honest), the trust chain will name at least one malicious entity.

To summarize: we eliminate dependency on future loads by destroying the old private key; but the past loads (on which a given version depends) participate in the chain for that version.

7.3.3   The OA Manager

Since we do not know a priori what device applications will be doing, we felt that application key pairs needed to be created and used at the application's discretion. Within our software architecture, Layer 2 should do this work—since it is easier to provide these services at run-time instead of reboot, and the Layer 1 protected memory is locked away before Layer 2 and Layer 3 run.

This *OA Manager* component in Layer 2 will wield a key pair generated and certified by Layer 1, and will then generate and certify key pairs at the request of Layer 3. This approach follows our guidelines: the ratchet locks (Section 7.3.1) ensure that the Layer 1 cannot depend on the OA Manager; the OA Manager depends on Layer 1, but Layer 1 creates and is named in its chain.

When requesting a key pair, the application specifies whether it should live as long as that Layer 3 epoch or that Layer 3 configuration. The OA Manager

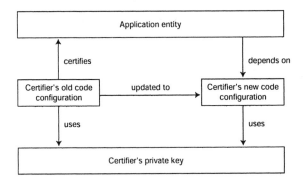

Figure 7.6. If the certifier outlives its own code change, then the application can incur a dependency not named in its chain.

will indicate this in the certificate; in conspiracy with Layer 1, the manager will also enforce this lifetime, by using our special BBRAM regions to see that the private key is zeroized when the lifetime ends.

These certificates also indicate that said key pair belongs to an application, and also include a field chosen by the application. (A straightforward extension of our trust calculus would thus distinguish between owning and trusting a key pair for certification purposes, and owning and trusting a key pair for the application-specified purpose—the last link.)

How long should the OA Manager key pair live? To keep the chain linear, we decided to have Layer 1 generate and destroy the OA Manager key pair (e.g., instead of adding a second horizontal path between successive versions of the OA Manager key pairs). The question then arises of when the OA Manager key pair should be created and destroyed.

We discuss some false starts.

As Section 7.1.7 discussed, the interaction of certifier and certified lifetimes causes trouble.

If the OA Manager outlived the Layer 2 configuration, then our certification scheme *cannot be both sound and complete.* Figure 7.6 shows an example. Suppose Layer 2 is updated from B_1 to B_2 while preserving the OA key pair k_B. Application C_1 depends on the new version B_2, but its chain only names B_1. We violate the theorem: the scheme cannot be sound and complete.

If the OA Manager outlives the Layer 3 epoch then we also have trouble. Figure 7.7 shows an example. Suppose application C_1 is replaced by application C_2 but the OA Manager retains the same key. If a relying party worries that C_2 may penetrate the OS, then C_1 may incur a dependency on C_2—even though the C_1 chain does not name C_2.

Our final design avoided these problems by having the Layer 2 OA Manager live *exactly* as long as the Layer 3 configuration. Using the protected BBRAM

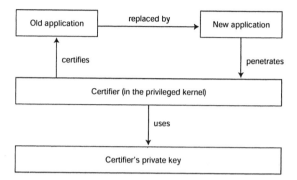

Figure 7.7. If the certifier outlives the application, then the old application can incur a dependency not named in its chain; for example, if a new, untrusted application might manage to penetrate the OS barrier.

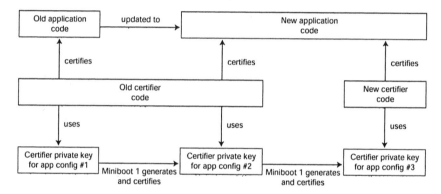

Figure 7.8. We insure that chains name dependency by having Layer 1 generate a new OA Manager key pair with each change to code the application depends on.

regions, we ensure that upon any change to the Layer 3 configuration, Layer 1 destroys the old OA Manager private key, generates a new key pair, and certifies it to belong to the new OA Manager for the new Layer 3 configuration. If the new configuration was due to a Layer 1 reload, then the old Layer 1 signs a transition certificate which signs the new OA Manager key pair. This approach ensures that the trust chain names the dependency set for Layer 3 configurations—even if dependency is extended to include penetration of the OS/application barrier. Figure 7.8 sketches this structure.

(As noted earlier, the private halves of any Layer 3 configuration key pairs will also be destroyed; if this configuration change does not preserve the Layer 3 epoch, those private keys are destroyed as well.)

7.3.4   Naming

We already discussed the naming formats for the initial device certificate and the transition certificates. The OA Manager certificate names the Layer 1 certificate that signed it, this particular device, and and names the software entities (again, via identity of the owner of this layer, name they gave to this code, revision number they gave to this code, SHA-1 hash of this code, when the current layer epoch started, when the current layer configuration started, etc.). in both Layer 2 and Layer 3. An application certificate names the OA Manager certificate that signed it (which thus names the current Layer 3 epoch and the software present in this Layer 3 configuration), whether this certificate lives for an epoch or just a configuration, and the arbitrary data field given by the application.

Trusting an epoch-entity requires, by definition, gambling that future secret-preserving code changes will be trustworthy. To make this more reasonable, we include code owner information (so that the relying party can know whose judgment they are trusting). To accommodate parties who chose to trust epochs to later change their minds, note that we also ensure that a Layer 3 epoch certificate (say, for epoch E) still names the Layer 3 configuration (say, C_1) in which it began existence. If, in some later Layer 3 configuration C_k within that same epoch, the relying party decides that it wants to examine the individual configurations to determine whether an untrusted version was present, it can do that by examining the trust chain for C_k and the sequence of OA Manager certificates from C_1 to C_k. An untrusted Layer 1 will be revealed in the Layer 1 part of the chain; otherwise, the sequence of OA Manager certificates will have correct information, revealing the presence of any untrusted Layer 2 or Layer 3 version.

In a sense, a relying party exercising this "right of retroactive paranoia" begins with a trust set that treats configuration within an epoch in the same equivalence class, but then relaxes this assumption.

7.3.5   Summary

As noted earlier, the trust chain for the current Layer 1 version starts with the certificate the factory root signed for the first version of Layer 1 in the card, followed by the sequence of transition certificates for each subsequent version of Layer 1 installed. The trust chain for the OA Manager appends the OA Manager certificate, signed by the version of Layer 1 active when that Layer 3 configuration began, and providing full identification for the current Layer 2 and Layer 3 configurations and epochs. The trust chain for a Layer 3 key pair appends the certificate from the OA Manager who created it.

Our design thus constitutes a trust-set scheme that is sound and complete for the dependency function we felt was appropriate, for any legitimate trust set. A certificate for an OA Manager key pair names exactly those configuration entities (including Layer 3, in case one does not trust the OS protections) that correct use of the Manager's private key depends on. A certificate for a configuration-length application key pair names exactly those configuration entities it depends on.

A certificate for an epoch-length application key pair names exactly those epoch-entities it depends on; should the relying party later decide to not trust a particular Layer 3 configuration, a method exists, as sketched above, to shift to a configuration entity and determine if the untrusted configuration was present.

7.3.6   Implementation

Full support for OA shipped with all Model 2 family devices and the CP/Q++ embedded operating system.

Implementation required some additional design decisions. To accommodate small developers (Section 7.1.3), we decided to have the OA Manager retain all Layer 3 private keys and wield them on the application's behalf; consequently, a party who trusts the penetration-resistance of a particular Layer 2 can thus trust that the key was at least used within that application on an untampered device. Another design decision resulted from the insistence of an experienced application architect that users and developers will not pay attention to details of certificate paths; to mitigate this risk, we do not provide a "verify this chain" service—applications must explicitly walk the chain. We also gave different families of cards different factory roots, to encourage relying parties to make a conscious decision about the root they choose.

A few aspects of the implementation proved challenging. One aspect was the fact that the design required two APIs: one between Layer 1 and Layer 2, and another between Layer 2 and the application. Another aspect was finding places to store keys. We extended the limited area in BBRAM by storing a MAC key and a TDES encryption key in each protected region, and storing the ciphertext for new material wherever we could: during a code-change, that region's FLASH segment; during application run-time, in the Layer 2-provided PPD data storage service. Another interesting aspect was the multiplicity of keys and identities added when extending the Layer 1 transition engine to perform the appropriate generations and certifications. For example, if we decide to accept a new Layer 1 load, we now also need to generate a new OA Manager key pair, and certify it *with the new Layer 1 key pair* as additional elements of this atomic change. Our code thus needed two passes before commitment: one to determine everyone's names should the change succeed, and another to then use these names in the construction of new certificates.

As has been noted elsewhere [DLP+ 01], we regret the design decisions to use our own certificate format, and the fact that the device has no form of secure time (e.g., Layer 3 can always change the clock). Naming the configuration and epoch entities was challenging, particularly since the initial architecture was designed in terms of parameters such as code version and owner, and a precise notion of "entity" only emerged later.

7.4 Further Reading

This chapter is based in part on my *ESORICS* and *IJIS* papers [Smi02].

The details of the software configuration and epochs that this outbound authentication scheme provides may still not be sufficient for an external relying party to make a trust judgment, if the the relying party does not know how to evaluate whether the platform so configured is trustworthy for an application. Chapter 9 and Chapter 11 present approaches my students and I developed to allow an external level of indirection. particularly in the context of SSL Web sessions: the relying party delegates this evaluation to a CA who indicates the appropriate attributes in the certificate it signs for the TCP. Very recently, a generalized version of this idea—*property-based attestation*—has appeared in the literature [SS04].

Chapter 8

VALIDATION

The notion of a "trusted computing platform" centers, by definition, on *trust*. The appropriate stakeholders need to be able to trust the computation this platform carries out. However, as Section 1.1 introduced, the concept of "trust" is more nuanced than the naive way we in the security community often use the term. We need to worry about exactly what it is the stakeholder wishes to trust the platform to do. We need to worry about whether the platform is in fact worthy of that trust. Finally, we also need to worry about how it is the stakeholder knows that the platform is in fact worthy of this trust.

A large part of the value of the IBM 4758 architecture—described in Chapter 6 and Chapter 7—was that it was not just idle speculation. Rather, this work yielded a real product in the real world. Stakeholders included potential customers who needed to know why they should believe that our TCP was in fact trustworthy—and what this meant. In my security analysis work before IBM, I would continually tell clients "never trust a vendor." Suddenly I became one. Why should my former clients believe me?

To make the case to the stakeholders, we needed to find an external, independent entity to validate our TCP. Performing this validation against a standard set of criteria would give it more credibility than an ad hoc review. At the time, the *Federal Information Process Standard (FIPS) 140-1* seemed the standard (and validation process) most appropriate to our platform. FIPS 140-1 addressed security for cryptographic modules. Although cryptographic modules are not quite the same thing as a TCP, but FIPS 140-1 did talk about physically secure devices that did computation. Furthermore, an additional motivation existed: FIPS 140-1 had several levels, and no module had ever been validated at Level 4, the highest security level. By my recollection, we told the business units that we were aiming for Level 3—but went for Level 4 anyway. We succeeded in earning the world's first validation at FIPS 140-1 Level 4, and (to date) no other general-purpose programmable device has even equaled this level.

This chapter discusses this experience.

- Section 8.1 presents the background of the FIPS validation process.

- Section 8.2 presents our validation strategy.

- Section 8.3 formalizes the security properties I deemed important for our architecture;

- Section 8.4 discusses we formally verified that our architecture had these properties.

- Section 8.5 discusses the other aspects of the validation.

- Section 8.6 provides some overall reflections on the validation process.

8.1   The Validation Process

8.1.1   Evolution

The Orange Book. Perhaps the most well-known set of computer security standards is 1985's *Trusted Computer System Evaluation Criteria (TCSEC)*, the U.S. Department of Defense's rules for computing systems [Dep85]. (This document—and the rules and world view it puts forth—is commonly known as the *Orange Book*, after the color of its cover.) The Orange Book intended to give users a "yardstick" to measure security, to guide manufactures in designing and producing secure systems, and to guide procurement officers who need to decide what computers to buy for DoD applications. For the DoD, the security goal was making sure that classified data does not fall into the wrong hands. The Orange Book focuses both on *features*—what the system must do—as well as *assurance*—why someone should believe it does that. It set up a series of ratings, increasing in required functionality and assurance. A system was also put in place whereby vendors could submit systems to receive ratings, according to this yardstick.

As an aside, a number of apparently synonymous terms might describe the process of earning such a rating: "validation," "certification," "evaluation," etc. The reader should note two things. First, some subcommunities are very persnickety about which term applies to their particular standards process; so if the audience looks aghast when one uses one of these synonyms, then try a different one. Second, marketing literature will occasionally use a term like "compliant"; this means that in the judgment of someone such as a marketing executive, the product might satisfy the criteria of the standard. If one is the audience for such a usage, interpret it to carry the rigor it deserves.

Physical Security. As noted, the DoD focused on protecting classified data within a larger computing system. As a result, the Orange Book developed criteria—such as for labeled data and sanitization of re-used objects—for supporting sensitive data and operations within a larger software environment.

However, these criteria were not directly relevant to trusted computing platforms intended to provide security against adversaries with direct physical access. As a consequence, in 1990, Steve Weingart (behind the physical security designs for μABYSS and the 4758), Steve White (behind ABYSS and Citadel) and colleagues proposed evaluation criteria for physical security [WWAD90].

The proposal focused not as much on defensive techniques as it did on the difficulty of successful attacks. For physical security, the proposal laid out a series of six levels (Level 1 through Level 6) of increasing security, over four different categories. A system's overall rating was its minimum rating in these four categories. At the higher levels, the proposal required the evaluators to try attacks not anticipated by the designers. This proposal also made a point of considering both the platform as well as the environment in which the platform was used.

8.1.2 FIPS 140-1

In 1994, the U.S. Government followed up with FIPS 140-1 [Nat94]. As noted earlier, this standard was intended to address cryptographic modules: devices that performed cryptography (and perhaps other computation), and ensured the confidentiality of cryptographic keys—as well as the integrity of cryptographic algorithms—against an adversary with direct physical access. FIPS 140-1 reduced the six levels of the Weingart et al classification to four, across a wider set of categories. As in the earlier classification scheme, a module's overall rating is the minimum of its ratings in the various categories.

Given that "cryptographic modules" can show up in a variety of forms in the information infrastructure, FIPS 140-1 had an additional axis. Besides "level" and "category," FIPS 140-1 also had different rules depending on what type of module it was—ranging from software-only to single-chip to multi-chip. The rules for software modules tried to establish equivalences to the physical security of the hardware modules by requiring Orange Book validations for the larger software environment, but (as Peter Gutmann observed [Gut04, Section 7.1.2]), this led to "impedance mismatches": relatively insecure commercial operating systems being ranked equivalent to relatively high-security physical modules.

Perhaps as a consequence of eliminating two levels, a substantial gap exists between FIPS 140-1 Level 3 and FIPS 140-1 Level 4. On the physical level, the criteria move from withstanding a few suite of prescribed tests to virtual impenetrability; on the software end, the criteria move beyond substantial documentation to a complete formal mathematical model and (within that model) formal proof of security.

8.1.3   The Process

With the FIPS 140-1, an independent laboratory carried out the validation, under the auspices of the U.S. and Canadian governments. This process could involve six months or more of interaction with the laboratory. The fact that the vendor paid the bill—plus the fact that this interaction took design and implementation staff away from product preparation, and potentially delayed shipment—made validation a substantial process, not to be undertaken lightly (and forbidding even for larger vendors).

Determining exactly what the validation involves also was surprisingly complex. One starts with the standard itself, which (at 55 pages) is almost manageable. However, the *Derived Test Requirements* [HMMW95], over twice as long, provides much more specific details regarding the actual testing. Furthermore, much as law consists of the original legislation filtered through the prism of case law, FIPS 140-1 consisted of these documents filtered through the *online implementation guidance.* All of these items referred to other FIPS and ANSI standards for specific cryptographic algorithm specifications. The fact that the original standard document fans out to all these other documents—and the fact that the online guidance and (occasionally) the cited standards would continually evolve and change during the non-trivial interval in which a module underwent validation—made this process even more challenging.

8.2   Validation Strategy

Our device is a secure coprocessor *platform.* As Chapter 6 discussed, a set of external authorities can conspire to configure a particular platform with a certain software stack. Each layer in this stack can then execute and accumulate state, as long as the device remains untampered and the environment supporting that layer remains in a state that the authority over that layer deemed trustworthy a priori. Thus, what we want to validate is that this works: the platform is a physically secure package guarded by secure bootstrap/configuration control software. "Trust us" is not good enough, and a validated platform would make it easy for follow-on applications to get validated.

A priori, we decided to separate the software validation effort from the hardware validation effort. This section describes our initial plan to validate the software. (Section 8.5 below discusses hardware.)

The software validation required several different components:

- The validation process required a *security policy.* In textbook and classroom settings, this is a chart indicating who can do what to whom when; the policy has the implicit goal of being small enough for a human can analyze it and determine what it is the system does, and whether it makes sense. In development settings, the security policy also becomes something one can tack on the wall (at least, that's what I did) and use to guide implementation decisions.

However, the FIPS validation process required the policy to be in a special (and rather lengthy) format; this ended up being different from the textbook policy that guided the implementation.

- The validation process required a *finite state machine (FSM)* of the system software. In software development, a programmer is used to thinking of the system as being in some particular condition; code transforms the system condition, based on the current condition and external inputs. These notions extend to formal modeling: the system is in some state, and processing and events transform that state.

  However, in the established tradition for FIPS validation, the "state" is not the system conditions that get changed, but the actual transition function itself. This counter-intuitive terminology required us to think of some other term for the system state that gets transformed; it also led to continual misunderstandings from our research colleagues. (E.g., "what's so hard about verifying a system with fewer than 1000 states?")

  The FSM appeared to have two roles in the validation process: both to serve as a guide for the human validator to understanding the code, as well as to serve as a foundation for formal analysis.

- The validation process required a *formal mathematical model* that describes how the system behaves. Within this model, we needed to specify the *invariants* that described the important security properties, and then verify that system behavior (within this model) preserved these invariants.

- The validation process also required extensive documentation: source code, annotated to show correspondence to the FSM; diagrams of the FSM; discussion of each state; and exhaustive state transition tables.

The validation process did not require mechanical verification that the security invariants held in the formal mode. However, given the advances in near-industrial strength automated formal methods (see [CW96] for a survey of the state of the art then), we decided to use automated methods: both to gain increased assurance that the invariants in fact held, and also (given the "case law" flavor of FIPS, and the fact that we were blazing new ground with Level 4) to set a strong precedent.

Automated formal methods come in two main varieties.

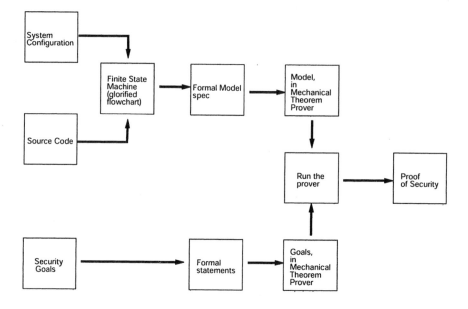

Figure 8.1. The formal verification process, as we envisioned it before we started.

- *Model checkers* search a state space for violations of an invariant. A negative result gives details about how to reach a counterexample; a positive result simply says "no violations."

- *Theorem provers* instead produce a proof that the model satisfies the security property. A positive result yields a proof that (in theory) a human could read and check; a negative result simply says "could not prove it."

Since we believed the system worked, and since the idea of a checkable proof of security seemed appealing, we chose the latter route. Since we had access to an installation of it (and an experience pool) within IBM, we chose to use the ACL2 theorem prover [KM97].

Figure 8.1 shows the process we planned. As designer and developer, I would take my knowledge of the system configuration and software and produce the finite state machine. Working with Vernon Austel, we would abstract this description into the formal model, and then embed it in the language of the theorem prover. We would then take the security goals of the system, abstract them into formal statements, and embed them in the language of the theorem prover. We would then just turn the crank, and produce the proof of security.

Needless to say, it did not quite work out this way. Section 8.4 will expand on what happened.

8.3 Formalizing Security Properties

Our validation effort depicts our TCP as an armored room with a burly guard at the door. The software component of the validation focuses on this burly guard: permanent Miniboot 0 and rewritable Miniboot 1, which control device security. In terms of a classic policy, these entities provide services to the various external code authorities, who can issue commands during the time Miniboot runs to do things like load code, establish child officers, and help do an initial load of the child's code. During an ordinary boot (with no configuration-changing commands), a user can request a signed statement about what's in the card, and (implicitly) invoke Layer 2 and Layer 3 code.

Given the FIPS notion that "officers" are the privileged entities that invoke privileged module operations, the code "Authorities" from Chapter 6 turned into "Officers." (We also had to push the boundary somewhat, in that, in general, our Officers communicated their commands via signed requests—with considerable space and time separating the signing of the command at the Officer site from the playing of the command to the module.)

The Miniboot code also cleans up after various other events: resets occurring during its operation, detection of component failures, and actual tamper. This reality quickly taught us that our a priori separation of hardware from software was misguided; the software formal model needed to include hooks to represent the effect these real-world hardware events would have on the operation of the software.

Throughout operation, Miniboot (and hardware) manipulates fields such as:

- the programs and the public keys in each FLASH segment;

- the identity of each officer;

- the state of each FLASH segment;

- the state of the overall device initialization;

- and the state of the secrets for each layer.

These became the foundations of our "system condition" space (what we would have called "state," except the FIPS tradition already used that term for the transformation functions).

One our main goals with this formal verification was to establish *safety*. We know from testing that the device does what it is supposed to do under ordinary conditions. However, given the many separate pieces, the split between Miniboot 0 and Miniboot 1, and various disasters and failures that may occur, we were much more concerned about verifying that the device not do something insecure under some bizarre, hard-to-test scenario.

To do this, we define a sequence of primary invariants. We then mechanically prove that, when the device starts in in the configuration resulting from "Factory

Initialize" and we do not load something other than Miniboot 1 into Layer ‵, that these invariants remain true.

Informal Invariants. Our security software (in conspiracy with the hardware) makes two central commitments.

- Only the current Officer over Layer N in a device gets to control that Layer N 's software environment, in cooperation with his parent Officers.

- Only that layer gets to see the secrets it accumulates, despite the various bad things that may happen.

We need to transform these commitments into more formal invariants.

8.3.1 Building Blocks

We need some formal notation to use as building blocks here.

First, recall that (in FIPS lingo) a *state* is a period of transformation: when the hardware is executing some chunk of code. We thus use the term *device configuration* to describe what the rest of us would call "state:" the state of the device, its critical fields and parameters, etc.

Execution States. We need to be able to talk about the software execution states corresponding to each layer of software, and to order the execution states by layer and reachability, in order to say things like "Property X must be true, if we ever get beyond State S."

Before boot time, the device hardware may act. (We put reaction to reset and tamper in this category too.) After that, the software is executing. Let *Exec_State* be the set of reachable states, in which the system is executing. We partition this into the sets *Exec_State_0* through *Exec_State_4*, for each layer's period during boot (plus runtime, after the application initializes). We then define a partial order on these states that respects reachability order—and also follows the striations of this partition.

Actors. We need to describe what the device is, and who controls it, up some layer p. We define $Actor_p(C)$ to be the tuple of parameters describing the state of the Layer p configuration, in device configuration C. An *actor* is either such a p-actor, or the special actor *Hardware*, to denote actions taken by the hardware itself.

For state S, define:

$$StatesActor_S(C) \quad = \quad \begin{array}{ll} Actor_p(C) & \text{if } S \in Exec\_State\_p \\ Hardware & \text{if } S \notin Exec\_State \end{array}$$

$StatesActor_S(C)$ describes who is acting, when the device is in state S.

Bad Hardware. We define the predicate $BadHW_S$ (C) to be true when configuration C is unsafe for state S (e.g., due to failure of some critical memory component).

Healthy Environment. For a software layer to run fully and correctly, the device must have some minimum level of health beyond simply not having bad hardware. We define the predicate $Healthy\_Env_p$ (C) to be true when configuration C has appropriate hardware and sufficient initialization to run Layer p in state S.

8.3.2 Easy Invariants

Some of the invariants we proved were fairly straightforward.

Clean-up. As we discussed in Chapter 6, the different software layers can be in different states (e.g., "Unowned," "Runnable," etc.). The device itself can have different levels of initialization (e.g., whether it has the Miniboot 0 secrets, whether Miniboot 1 has a certified key pair, etc.). During design, I crafted a couple of straightforward matrices showing what combinations were legal and what were disallowed; during implementation, I tried to ensure that the Miniboot code would ensure that the disallowed combinations were not reachable.

Almost as an exercise, the first invariant we verified was: does this clean-up work? Under the various reachable states (coupled with hardware failures and interruptions), do we ever end up in the disallowed regions? The answer, fortunately, was "no."

Safe Access. We then established that lower-privileged software cannot access the secrets belonging to higher-privileged software. That is: if S $\in$ $Exec\_State\_p$ then $ratchet \geq$ p. This is also fairly easy to show: after initialization, you cannot jump across layers without the ratchet increasing.

Safe Execution. We also showed that the device cannot execute in a state where it depends on hardware that has failed. So, for any reachable S ,C , we show that $BadHW_S$ (C) = *false*.

(Note that the our failure model makes the simplifying assumption that physical failures stop the device. As a consequence, this model does not address the window between the testing for failed hardware, and—within the *same* reset cycle—executing code that depends on that hardware.)

8.3.3 Controlling Code

The Miniboot security policy gives lots of options about which Officer can change what, under what conditions. However, with our "safe control" invari-

If Layer p here...   ...does not dominate Layer p here

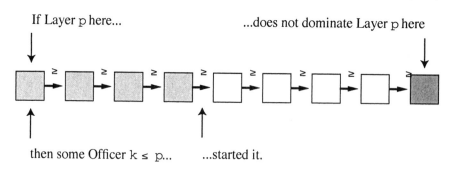

then some Officer k ≤ p...   ...started it.

Figure 8.2. The "safe control" invariant asserts that, if a layer changes in a way other than decay, then that layer's officer—or a superior—started the process (which could, conceivably, have included changing officers at some point).

ant, we describe the overall goal of this table—by proving it, we verify that the policy makes sense.

Given some configuration C , *who* can change layer p in the device? Only Officer k, for k ≤ p—using the key they had in C . So if you trust yourself and your superior officers *now* to behave themselves, you can trust that your layer will remain unmolested.

Some of the wrinkles here that we needed to address included the fact that a Layer may "decay" due to various types of failures, and the fact that Officer k for this device at some future point in time might not necessarily be the same party as Officer k is now.

We needed to define a partial order on the status of each layer, in order to distinguish between decay, and other types of change. We then formalized (and verified) the statement that Figure 8.2 sketches: if a sequence of Layer p configurations exists where the initial Layer p status does not dominate the final one, then some Officer k (for k ≤ p) issued some successful configuration-changing command, using their identity and keys that existed in the initial configuration in this sequence.

8.3.4 Keeping Secrets

Officer p's program runs and accumulates secrets. However, various bad things can happen that cause the device to stop being a safe place for these secrets. These bad things include:

- hardware attacks

- hardware failures

- changes to the contents of layer k (for k < p) that Officer p did not trust

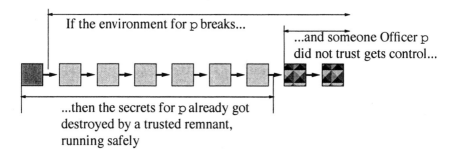

Figure 8.3. The "safe zeroization" invariant asserts that, if the environment for a layer stops being trusted and safe for that layer, then its secrets will be destroyed by a trusted remnant, running safely, before any potential adversary (or untrusted software) gets control.

- changes to the contents of layer k that do not necessarily involve the participation of Officer p

We need one clean way to say "one's secrets are safe." The problem in *stating* this invariant is that "safe" means different things, depending on the bad thing that happens:

- Hardware attacks cause instant zeroization.

- Untrusted changes to parents cause clearing now (during the change) or at the next boot (before anyone gets to run).

  Furthermore, if the parent in question was Miniboot 1, then the secrets had better be cleared by the old, trusted Miniboot 1 and/or the permanent Miniboot 0 *before* the new Miniboot 1 gets a chance to run.

- Untrusted changes to contents of layer p should cause clearing before layer p gets a chance to run again.

- Certain POST failures cause the secrets to remain intact, but the device then never goes beyond POST.

The problem in *proving* this invariant is that, because "bad things" do not necessarily result in instant clean-up, additional bad things may occur before the device has finished responding to the first one.

The "safe zeroization" assertion we developed is *temporal:*

- if a bad thing happens *now,*

- and an adversary might see the secrets *later,*

- then *sometime in between,* the secrets have been destroyed.

Figure 8.2 sketches this invariant.

To express this, we assume we are given p and a starting configuration S_1, C_1. We then define three new auxiliary variables that "remember" the relevant events:

- **Has the environment broken for** p**?** We define the *Env_Broken*$_p$ predicate to start out false, but become true when something happens that makes *Healthy_Env*$_p$ fail; when some $k \leq p$ has an accepted emergency burn (since these can never be trusted); when some $k < p$ has an ordinary burn that is not trusted by Officer p; when some $k \leq p$ surrenders; and when some $k \leq p$ becomes not runnable.

- **Who is trusted to do the clean-up?** We define *Trusted_Remnant*$_p$ to be the subset of the hardware and Layers that Officer p trusts. Initially, this is everything up to p; however, various configuration changes, failures, and tamper can reduce this set.

- **Has the clean-up happened yet?** We define the predicate *Secrets_Destroyed*$_p$ to become true when the Layer p secrets are destroyed.

We then can state the invariant. Suppose the device is in some state S and configuration C. Layer p is runnable, and the configuration and satisfies *Healthy_Env*$_p$. From then on, if *Env_Broken*$_p$ becomes true but the current layer acting is not in *Trusted_Remnant*$_p$, then *Secrets_Destroyed*$_p$ is true.

8.4   Formal Verification

Figure 8.1 above sketched the formal verification process we initially planned. Figure 8.4 sketches what actually happened: the FSM and FM merged, and we required much iteration on the mechanics to get the proofs to work out.

One of the surprises was the difficulty in abstracting from the finite state machine to the formal model. The formal model was supposed to describe what the device does, and the FIPS documentation rules essentially required a deterministic, total function for each FSM state. Since non-trivial abstraction from the FSM to the FM was not working, we ended up merging the two: the formal model was built directly from the FSM.

Other issues arose in defining the functions for each state. Since device behavior is not determined by software alone, we needed to add hardware events to our model, to describe how hardware passes control to software, and to describe how software reacts to hardware events. Describing what the device "does" depends on the level of abstraction of the view—e.g., the particular Miniboot 1 signature verification code executing right now, or the fact that if we drill a hole, the device will zeroize. To accommodate these layers of abstraction, we built a hierarchy of finite state machines/models, where a state in one can expand into a child FSM. This allows us to have one function per state, while also encapsulating functionality common to many states in one place

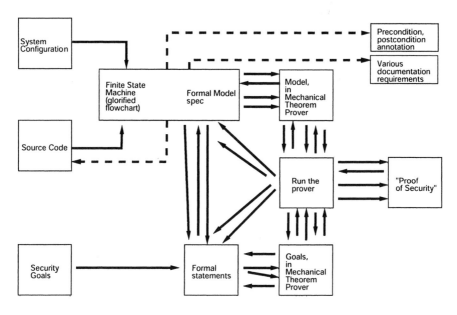

Figure 8.4. The formal verification process, as it actually happened.

(explicitly linked via a *Parent_Trap* hook). Standard wisdom in programming is to put common functionality in one place. The same wisdom applies when building a formal model.

Sometimes, our code states could embody a natural reaction to some external event (such as a failure or some Miniboot command); other times, the natural granularity of the software structure required successive states with no natural trigger event. Thus, we developed an event set that included the generic "tick," and a driver function that steps through the model, taking into account events and parents.

I obtained the FSM states by partitioning the source code into small chunks, and associating atomic transformations with each state. The code was written so stateful changes are atomic, but we still required splitting states here and there. These software states varied significantly in complexity; some configuration clean-up states each represent just a test or two, but FLASH burning can take lots of code. To structure the FSM cleanly, we even had switch states that did not change the configuration at all. Other war stories included carefully constructing component sets to cover every scenario formally, and developing *structural validity* predicates to indicate which combination of tuples were in fact legal.

Another surprising challenge was keeping the documentation synchronized. Figure 8.5 shows the various documentation elements and their linkages. With

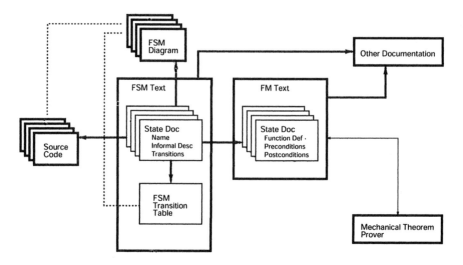

Figure 8.5. To keep the validation documentation synchronized despite the continual fluidity, I developed some customized tools.

the continual fluidity induced by the various tuning and refining necessary to get the formal modeling to succeed, it was easy to lose correlation. If we changed one element, how many places in how many different files (in how many different formats) need to be touched? To address this, I used symbolic references everywhere. I developed custom LATEX macros (and post-processing code, in C) to create the transition tables and formal model documentation, and update the FSM diagrams and source code annotation, from master changes in the state documentation.

From all of this, we learned that the software works (at least, within its abstraction as this model). However, it was also good that I designed it for modeling eventually; had we not been aiming for that goal all along, it is not clear how feasible this process would have been. We learned that, as in programming, concise expression is important. In the model, it was good to put common structure in one place; in the documentation, it was good to have automated tools to keep things correlated. We learned that time and temporal order was critical in the security assertions we needed to approve (and that our theorem prover did not like this). We also learned that the a priori division of hardware and software was not always appropriate.

8.5   Other Validation Tasks

The validation process also required other tasks.

Hardware. One of the largest tasks was showing that the physical security withstood the Level 4 scrutiny. Some of the issues here included quantifying exactly what "no undetectable penetration" meant; clarification emerged regarding minimum hole sizes for conductive and non-conductive probes that must trigger tamper. (It was also rumored that, due to the tamper-resistant nature of our packaging, the validation lab kept snapping drill bits, and one our units under test ended up looking like a "porcupine.")

Algorithms. We needed to prepare excruciating documentation for all cryptographic algorithms, and have the device test them at run time. We had hardware support for many, and the POST code prepared by the hardware engineers tested this; however, as yet another instance of the wrong a priori division of hardware and software, we had to re-test these in software, since validation required looking at the entire stack—software and hardware—as seen by the party using the cryptography.

For "approved" algorithms (basically, algorithms for which a FIPS or other appropriate standard existed), we also needed to go through validation tests to prove that our implementation indeed conformed to the standard. This task was trickier than it needed to be, due to formatting and other ambiguities. We also had problems because the de facto signature standard was RSA—but this was still not approved at the time of validation. As a consequence, we needed to add DSA support and options for all Miniboot places that used signatures, precisely so customers could then proceed not to use it.

Randomness. Obtaining approval for our randomness sources was also challenging. The rules that the random number generator must satisfy statistical tests at power up and continuous tests throughout operation, and that (for keys) hardware bits must go through a FIPS-approved pseudo-random number generator. We added statistical and continuous tests on the hardware RNG, and obtained RSA and DSA keys through the DSA PRNG, and obtained DES keys through a DES-based PRNG.

However, even though our cryptographers could establish that DES was sufficiently pseudo-randomizing, the validation lab would not accept this argument. So, we sent DES keys through the DSA PRNG too.

The validation lab then would not accept our statistical and continuous tests, since they applied to the hardware RNG, not to the PRNG. We pointed out that since the hardware RNG was working fine, and it was seeding the PRNG algorithm that the validators approved, and that our self-tests had shown the PRNG code had not changed. Wouldn't this be sufficient? Our argument was rejected. We added the extra set of tests—and found a bug! (It turns out that, due to all the changes in PRNGs, one of the contexts was not getting populated, and a PRNG context of zeros will yield a sequence that fails the tests.)

Operational Testing. The validation laboratory provided a list of items that we needed to demonstrate via operational testing. For each element in this list, we needed to interpret what it meant (for our device), verify that the device does it, submit a written plan for demonstrating it, and then demonstrate it for the validators.

This task forced me to prepare even more custom software. I needed to extend the test platform to support scripting and also to trigger all reachable errors. This testing requirement conflicted with good "belt and suspenders" defensive coding, since some failures were not reachable, because earlier code tests caught them. We needed to break Miniboot, in order to simulate FLASH failures; we needed to customize the hardware, to force zeroization and to break the hardware RNG.

In theory, these tasks are what one does for product development anyway; in practice, these tasks were an extra burden, due to the highly specialized nature of the operational testing requirements.

Follow-ons. Our Model 1 device earned the world's first FIPS 140-1 Level 4 validation.

Afterwards, we upgraded to Model 2 with more advanced hardware and with full support for the outbound authentication design in Chapter 7. We repeated the Level 4 validation for Model 2. We also carried out a Level 3 validation for our device, with the CP/Q++ software in Layer 2; the complexity of this legacy code made the formal modeling requirements of Level 4 not feasible.

8.6   Reflection

Overall, the existence of the FIPS 140-1 validation process and our passage through it served the purpose of helping to establish assurance that our TCP works as advertised. On the other hand, it took a great deal of time and expense, and consequently is beyond the reach of many developers. Validation requires resources and time that do not fit into the typical product lifecycle. These complaints were raised for the older Orange Book, and also for the newer Common Criteria (see below); how to reduce the barriers that seem endemic to formal security validation is an area of active research.

When preparing the follow-on standard, FIPS 140-2 [Nat01], NIST solicited comments. Numerous parties—both our team and others—raised various technical issues. The new standard addressed many of these: such as adding configuration management requirements for product lifecycle, and establishing some defense standards for newer side-channel and induced fault attacks that emerged during the life of 140-1 (recall Chapter 3). Another route that a TCP designer—or user—could chose is the *Common Criteria* [Com04], a newer international standards and validation system that emerged in the late 1990s. Departing from the Orange Book (which linked functionality to assurance in its ratings) and somewhat from FIPS 140-N (which provided some functional variation), the Common Criteria separates the security functional requirements from the *evaluation assurance level.* A *security target* describes the functionality and assurance for a particular device; a *protection profile* describes security requirements and assurance levels for a class of devices or systems. (However, the standard itself observes that some "technical physical aspects of IT security" are outside the scope of the Common Criteria, so FIPS 140-2 may be remain appropriate for TCPs.)

Neither FIPS 140-2 nor the Common Criteria address what I see as one of the fundamental problems with these standards: how to balance the details that describe what a particular device does with the generality necessary for an overworked manager to make the appropriate procurement decision. Even with FIPS 140-1, we had to wrestle to make a standard designed for crypto boxes fit our general-purpose programmable platform; the crypto library for Netscape looks very different from a postal meter, but FIPS 140-1 was used for both. On the other hand, the four levels, coupled with the varieties of module types and the fact that each module fit into a larger deployed system in a different way, was already too complicated. E.g., when the Bond attack on the CCA application surfaced [BA01], explaining that CCA was an unvalidated application living on top of a Level 3 validated OS layer living on a Level 4 module was overly complicated for some parties.

FIPS 140-N has not enough variations, and too many. The Common Criteria seems to go further in the "too many" category, but time will tell.

8.7 Further Reading

[SA98] gave a preliminary sketch of our formal modeling strategy; unfortunately, the promised more complete exposition never made it out of the draft stage. [SPWA99] gives an overview of our FIPS validation experience. Obtaining a good general picture of the Common Criteria from reading the standard itself is difficult; more concise (but perhaps slanted) introductions can be obtained from the commercial laboratories that carry out the validations.

Chapter 9

APPLICATION CASE STUDIES

Chapter 1 through Chapter 8 have toured the evolution of an armored, general-purpose secure coprocessor as a trusted computing platform. This chapter will now present case studies that used this TCP to solve distributed trust problems. Section 9.1 will review the basic building blocks. Section 9.2 will use this TCP to harden Web servers and their applications. Section 9.3 will use this TCP to bind an archive of sensitive information to the policy established for its use. Section 9.4 will use this TCP to allow clients to request and modify data stored at a server that learns nothing about the access patterns. Section 9.5 explores some other applications, and[1] Section 9.6 discusses some lessons learned from this experience.

9.1 Basic Building Blocks

Let's start by reviewing the computational model we're considering.

The applications in this chapter were all built around the IBM 4758, because that's what we had. However, we will generalize from the immediate reality of our prototype environment to a more general architectural model. Any system supporting these features could (in theory) support these applications. The limitations of the real platform sometimes shaped the application challenge; sometimes, the applications suggest some new architectural features.

Execution Environment. The TCP has secure non-volatile memory, zeroized upon tamper. The TCP has an execution environment—with volatile memory, CPU, and access to the secure non-volatile memory—that is also shielded from the adversary and zeroized upon tamper. The TCP has non-volatile program

---

[1]This is just a test, to see how the footnotes come out.

memory within its protected environment, but we do not assume that this memory is zeroized upon tamper.

Configuration. The TCP houses an onboard application entity. We assume that that the TCP family supports multiple developers and a broadcast distribution model (although more limiting assumptions would only make the "design problem" easier).

Outbound Authentication. The TCP can ensure its application entity has exclusive use of a private key, and can bind the corresponding public key to parameters sufficient to enable a remote relying party make a reasonable trust judgment about this entity.

Hardware. Some applications greatly benefit from the TCP having the ability to quickly transfer data across the secure boundary through a symmetric cryptography engine of sufficient strength (e.g., TDES or AES).

Limitations. We assume that the advantages of the secure environment come with a price: the TCP's internal computation environment (e.g., processor speed and memory size) will be small in comparison to standard desktops. Consequently, the TCP will likely function as a coprocessor, attached to a larger host machine that cannot be trusted against the adversary.

We also implicitly assume that the security comes with a monetary price as well; hence, most of the applications place the TCP at a server site, better able to afford the cost and amortize it over a larger pool of clients and transactions.

9.2   Hardened Web Servers
9.2.1   The Problem

Web Security. The Web is currently the main vehicle for information services in our infrastructure. To quickly review, the client (that is, the browser) sends a request to a remote server, which then responds with data (typically, in html format) which the browser renders for its human user. Without additional countermeasures, these exchanges take place in plaintext, exposing the data to scrutiny and potential modification by the adversary. Similarly, the browser user cannot be sure whether he has really established a connection to the intended server, or whether the adversary (by some DNS attack or other ruse) is impersonating the other end.

The *secure sockets layer (SSL)* has emerged as the standard way to address these risks. Like most protocols that make it into real world deployment, SSL has many variations. In its most typical use, a secure Web server possess a key pair. Only that Web server knows the private key (one hopes); one of the CAs whose public key is built into the browser's store of trust roots signs a certificate binding the corresponding public key to identifying information about the server. When the browser initiates an SSL connection, it does some basic validity and sanity checks on this certificate. The server proves knowledge of the private key; and the pair establish a session key and use this (via symmetric cryptography, with a MAC or hash) for the rest of their session.

If the server keeps its private key private (as well as the session key, and the parameters that led to its generation), then the use of the symmetric cryptography protects the data the parties exchange from modification and (plaintext) observation by the adversary. Each party might also reasonably conclude that the same entity is on the other end of the channel throughout the session—for otherwise, how would an adversary have known the keys?

If the browser and its user properly check the signature and data on the server's certificate at the start of the session, then, in theory the client might also reasonably conclude that the server is who they say they are. In practice, one might reasonably challenge whether these browser checks are reasonable, or whether clients notice if they fail, or whether the certificate even contained appropriate information in the first place. However, that is material for a different book.

Common SSL variations include servers that use a self-signed certificate (which sacrifices server authentication), servers that use a non-standard trust root (which requires that the relying parties install that root in their browsers), and installations where the client also has a certified key pair and proves knowledge of the private key as part of the initial handshake.

Armored Car to a Cardboard Box. However, a problem with using SSL to "solve" the Web security problem is that it only protects the tunnel between the browser and the server, and (in the standard instantiation) it only authenticates the server identity. Consequently, it becomes the proverbial[2] "armored car from a park bench to a cardboard box."

- What happens to the data once it gets to the server?

- What assurance does the client have that the server, even with that identity, is actually providing that service?

Indeed, I found an early inspiration for this problem when I was shopping for a bicycle component, and found the cheapest price at an online merchant I had never heard of. SSL assured me that my transaction could not be eavesdropped, and told me what the name of the merchant was. However, what I wanted to know was whether I could trust them to carry out this transaction (and not create

---

[2]Stressing the importance of matching security across components of a system, this proverb is a now-standard metaphor in security folklore.

a headache for me by misusing or accidentally leaking my credit card number); their name does not tell me this.

9.2.2   Using a TCP

A natural way to solve this problem is

- to bind a server's SSL private key not just to that server, but also to the application that server allegedly provides,

- and then to put this binding beyond the reach of anyone, even the operator of that server, to manipulate.

We could achieve this by welding the server end of the tunnel to the advertised application and moving both inside a TCP at the server site. The server operator obtains a TCP, then installs the application (augmented with the SSL tunnel end code). As part of its initialization, the application generates the SSL key pair, and uses the TCP's outbound authentication feature (Chapter 7) to prove to the satisfaction of a standard SSL CA that this TCP binds this public key to that application at that server. The CA then issues a special type of certificate, indicating this binding. (We might augment the browser to look for this special certificate type and indicate that.)

In Section 9.2.3 below, we revisit this CA-certification scheme, and in Chapter 11 we develop a more flexible approach.

Since we assumed that the TCP would be limited in computational power compared to standard machines, putting the entire server inside a TCP would not be feasible. Instead, the TCP with this application would function as a *trusted co-server*, providing an authenticated and perhaps secret shelter for this application, in way that client, server, and perhaps even other stakeholders could reasonably trust.

(The somewhat tongue-in-cheek original name of the project, *WebALPS*, emerged at lunch at IBM: "Web Applications with Lots of Privacy and Security.")

The participation of a trusted co-server in Web applications presents many application possibilities.

Credit Card Transaction Security. The current Web infrastructure provides secure transmission of a client's information to the server—but what happens there is anyone's guess.

For example, consider the credit-card information and transaction amount a client sends when he wishes to purchase something. An adversary who compromises the server (or a malicious server operator) can use this data to carry out lots of mischief:

- He can increase the amount of the transaction.

- He can retain the amount but repeat the transaction many times.

- He can use the credit card information to forge additional transactions.

This situation may significantly reduce the potential market for new e-merchants without a pre-established reputation. ("Ribo's Books has a cheaper price than well-known Amazon, but how do I know that unknown Ribo will neither steal nor accidentally divulge my credit card info?")

To solve this problem, the TCP-housed co-server can trap the credit card and transaction information, and then transmits it (within a safe cryptographic channel) to the acquirer's system. The credit card number data never appears in plaintext at the server site outside the TCP; the server operator or a penetrator has no opportunity to inflate the transaction amount; and (unlike SET) the client need not change the way she operates.

Nonrepudiation of Client Authentication. Without a public key infrastructure for citizens, most Web users are forced to use human-usable authenticators, such as userids and passwords. However, in the current infrastructure, these authenticators are exposed to the server of unknown integrity. As a consequence of this exposure, an adversary who compromises the server (or a malicious server operator) can impersonate this user at that site, and at any other site where the user has used these authenticators. This exposure also prevents legitimate server operators from being able to argue that it really was a particular client who opened a particular a session.

To solve this trust problem, the TCP-housed co-server can retain the the password, authenticate the client, then issue a signed receipt for th server that client properly authenticated for that session

Nonrepudiation of Client Activity. The current Web infrastructure prevents a server from being able to prove anything to a third party about the activity of an alleged Web client. For example, how can an insurance company taking an application from Alice over the Web later prove that Alice really answered that question that way? We would require a PKI for citizens, and perhaps a standard way to incorporate user signatures in Web form requests. (Such signed form support is not yet universally supported, unfortunately.)

For another way solve this trust problem that does not require these changes, the TCP-housed co-server can also issue a signed receipt for the entire transaction. "Alice not only authenticated correctly, but she issued a request of type X with parameters Y."

Nonrepudiation of Server Activity. The current Web infrastructure prevents a server from being able to prove anything to a third party about the activity of that server in an interaction. For example, consider trying to prove something

about the questions that generated the answers a client provided. Case law already exists that permits, in paper interactions, a client to alter a waiver before signing it—and if the service provider accepts the form without noticing the alteration, he is bound by it. For another example, some of my colleagues have reported difficulty in U.S. Government security clearance processes, because an answer to a question five years ago was not consistent with the current revision of that question.

To solve this trust problem, the TCP-housed trusted co-server can include, in its signed receipt for the transaction, the prompts and responses the server provided.

Taxes on E-Commerce.   The current Web infrastructure provides no acceptable means to balance the legitimate interests of a third party to accurately learn certain information about individual or collective Web interactions, with the privacy interests of the other participants.

For example, consider the problem of a government tax collection service trying to learn how much sales tax an e-merchant owes them for last month. Reporting all transactions to the government would be unacceptable to the merchant and customer for privacy concerns. Reporting only a total amount owed would be unacceptable to the government, since the figure would be unverifiable, and the merchant reporting this unverifiable figure would be motivated to understate it.

To solve this trust problem, the TCP-housed co-server can monitor the total tax owed by that merchant for the transactions that went through it (e.g., because of some other co-server application there), and report that authenticated total back to the government revenue agency. The agency can trust that the reported amount is correct; the merchant and customers can trust that the agency learns only what it is supposed to—and, in particular, is not learning details of transactions or identities of customers

Re-selling of Intellectual Property.   The current Web infrastructure provides no acceptable means for a third party who participates in an interaction indirectly, by licensing proprietary information to the server, to protect their legitimate interests. For example, a publisher who owns a large copyrighted image database might wish to make this available to a university library—but might worry that compromise of the university server will compromise the database.

To solve this trust problem, the TCP-housed co-server can receive a session key and licensing rules from the owner of the intellectual property. The owner would provide the intellectual property in ciphertext to the server; the co-server would decrypt the particular items being used, and ensure that whatever licensing/royalty/watermarking requirements were being enforced.

(Section 9.3 below expands on this idea.)

Privacy of Sensitive Web Activity. The current Web infrastructure provides no means for a server operator to plausibly deny that he is monitoring all client interactions. Similarly, the operator can cannot deny that an adversary who has compromised his machine is monitoring this data.

For one example, Consider people who wish to obtain sensitive literature — about health topics, for example, or about currently unfashionable politics. What prevents the server operator from learning of their activity, and acting in a manner (such as informing employers or health insurers) in way that would unjustly compromise user privacy?

To solve this trust problem for the data retrieval case, the TCP-housed co-server can implement some variation of an oblivious RAM algorithm that treats encrypted storage in the server's file system as the "RAM." The client, through the SSL-protected channel, makes her request to the co-server, which then retrieves the record via the algorithm, re-encrypts it, and returns it to the client. The server operator learns nothing[3] except the fact that a query has been made. Section 9.4 below will expands on this idea.

Correctness of Web Activity. The current Web infrastructure provides no means for a server operator to establish that he (or an adversary who has compromised his machine) has not otherwise altered or corrupted important correctness properties of the service.

For example, suppose an auction server provides a bulletin board service where customers can post "timestamped, anonymous, confidential" comments about participants and interactions. How can customers know that the anonymous posts came from bona fide customers, and that the timestamps are correct?

To solve this trust problem, we move the computation critical to the appropriate correctness properties from the server into the TCP-housed co-server — whose application program would need to advertise that it was performing these computations. This establishes that the trusted co-server witnesses that the alleged bona fide customers authenticated properly.

Enforcement of Logo Rules. The current Web infrastructure provides no effective means for a party to ensure that logos or endorsements appear only on the appropriate server pages.

For example, Dartmouth could establish a "Dartmouth-inspected" logo to endorse servers who have withstood penetration testing by specialists. However, any client who visits these pages can capture the logo and put it on any page, whether or not that site has withstood the testing.

---

[3] Strictly speaking, the server operator learns ciphertext; however, we implicitly assume that the cryptosystem is sufficiently strong that the adversary cannot distinguish it from random bits.

To solve this trust problem, the TCP-housed co-server could provide the logo information, when appropriate. Logos that do not appear in the portion of the browser window from an authenticated co-server-to-client channel are not legitimate. (Amir Herzberg has recently prototyped some browser work this vein [HG04].)

Safety of Downloadable Content. The current Web infrastructure provides no means for the client to ensure that executable content downloaded from a server is indeed safe. With the current continually penetrable state of consumer platforms, safety depends on the client themselves actually running the latest anti-virus software. Most consumers do not do this, leaving them at risk.

Moving the virus-checking computation (and the concomitant problem of maintaining the latest updates of virus signatures) to the server is more efficient—but how can clients know the server really carried this out?

To solve this trust problem, the TCP-housed co-server could run the anti-virus software with the latest signatures: either dynamically, as the co-server was feeding data back to the client, or offline (but then, when the co-server was feeding data back to the client, it would verify that it had indeed scanned this data earlier). Clients could then trust that content downloaded via this SSL-authenticated channel from the trusted co-server has been scanned.

Of course, without further optimizations, this application might tax the resources of the TCP considerably.

Authenticity of Downloadable Content. The current web infrastructure provides no easy means for the client to authenticate the origin of downloadable content. Posters of content can provide digital signatures, but then the client needs to explicitly obtain and verify the trust chain on each item.

Moving this verification computation (and the concomitant problem of maintaining the latest certificate revocation lists) to the server is more efficient—but how can clients know the server really carried this out?

To solve this trust problem, the TCP-housed co-server could itself verify the signatures of the posted content (using all the latest certificate revocation lists, etc.), and then include in the SSL-encrypted channel an assertion that this content had been verified, and the identity of its poster. The client Alice could then trust that content she downloaded via this SSL-encrypted channel from the trusted co-server did indeed originate with the alleged poster. (We can even save bandwidth here, since the client only need download the poster's identity—not his public key, signature, and appropriate certificates.)

Integrity of Server. The current Web infrastructure provides no means for the client to verify the integrity and site security of server machines.

For example, some servers may run on machines whose administrators who run hardened operating systems and/or engage in other good security practices, such as regular runs of a network security analyzer, or enhanced OS boot via secure hardware (e.g., as Section 4.3 discussed).

However, any site can claim to do this. How can a client know?

To solve this trust problem, the TCP-housed co-server can witness that the appropriate security tool (such as a network security analyzer or a particular hardware-directed secure boot technique) was applied to the host—perhaps because this tool was applied from the co-server itself, or from a companion trusted machine. The client Alice can trust such assertions she receives through the SSL-authenticated communication channel from the co-server to the client.

9.2.3   Implementation Experience

To see if the trusted co-server idea worked, we needed to implement it. At IBM, intern Naomura Itoi started working on it after he finished his Kerberos project (see Section 9.5.2 below) but failed to gain traction. When I moved to Dartmouth, my student Shan Jiang picked up the idea and prototyped it for his master's thesis [Jia01].

For the co-server platform, we used the IBM 4758 Model 2, with the CP/Q++ OS layer. For the server platform, we used a Linux desktop, with the Apache server (v1.3.14), which routes SSL work to the mod_ssl module (v2.7.1), which then uses OpenSSL (v0.9.6) to handle the cryptographic work.

Building the prototype required dissecting the SSL protocol to identify the minimum number of tasks we needed to move from the server to the co-server.

At an initial glance, we need to add eight messages (between the server and co-server) for each SSL handshake. The co-server tells the server its certificate (which it forwards to the client); the server forwards to the co-server the "Client Random" and "Server Random" messages; then the server passes on the client's "Encrypted Premaster Secret," "Change Cipher Spec" (to indicate the client is about to start using crypto), and encrypted "Finished" message. The co-server then responds with a "Change Cipher Spec" and its encrypted "Finished" message. However, we can reduce overhead by concatenating multiple server/co-server messages in one transmission and eliminating unnecessary ones. Figure 9.1 shows this revised handshake.

We then modified the server-side software stack to enable WebALPS and to carry out this modified handshake (as well as session re-use) with the co-server. Within the co-server, we used a hash table to track multiple concurrent sessions. However, in our prototype, we send every SSL request through the co-server. Having some sent through the co-server and some (of lesser sensitivity) through the server might improve performance, but the ability of browsers (let alone their users) to safely multiplex multiple SSL channels was not certain.

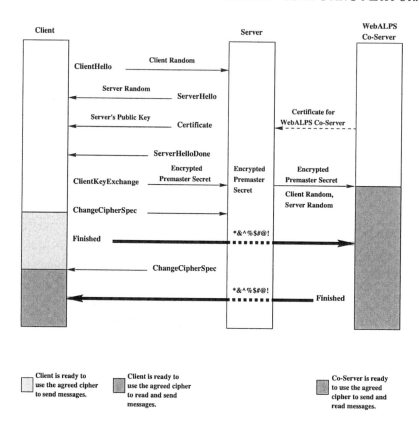

Figure 9.1. In our revised SSL handshake, the server forwards the key-derivation material to the co-server; only the co-server knows the keys to handle the encrypted, MAC'd traffic with the client. (Figure 4.6 from [Jia01].)

Our performance measurements sought to evaluate two things: how much the co-server slows things down, and whether the co-server approach can handle heavy workloads. The answers were "yes" and "yes," respectively.

We did the speed tests using the http_load tool, with a client that sent random requests for a 2KB file over a two-second interval. We measured served requests per second, connection time, and request time (once the connection was established), for an ordinary http server, an SSL server, and a co-server-enabled server (with a null internal application). Table 9.1 shows the results: WebALPS slows things down, but (as Table 9.2 shows) not by as much as SSL slows down http. (Some other cryptographic benchmarking work we have done [LS01] suggests that the CP/Q++ API to DES might be blamed for some of the slowdown.)

	Speed (requests/S)	Connection time (mS)	Request time (mS)
SSL and co-server	9.89	0.72	100.37
SSL alone	67.75	0.63	14.15
HTTP alone	858.80	0.18	0.91

Table 9.1. Comparing performance of an SSL server with a co-server to an SSL server without one, and to a an ordinary Web server. (Adapted from [Jia01, JSM01].)

	Speed (factor)	Connection time (factor)	Request time (factor)
Slowdown from adding SSL	11.68	2.46	14.6
Slowdown from adding a co-server	5.86	1.14	6.1

Table 9.2. Comparing the slowdown caused by adding a co-server to an SSL server, to the slowdown caused by adding SSL to to an ordinary Web server.(Adapted from [Jia01, JSM01].)

We did the workload test using the *WebBench* tool, which generates a representative e-commerce workload of 92% http requests and 8% https requests. We varied the tests over a number of clients and a number of threads per clients, and measured both requests-per-second and bytes-per-second. Figure 4 in [JSM01] shows the result: the performance is slower than traditional SSL, but things do scale.

Applications. With our prototype, the trusted co-server decrypts the ciphertext sent by the client, and encrypts the plaintext the server wishes to send back. The co-server thus has first crack at the plaintext the client sends, and last crack at the plaintext to be sent back; it can remove sensitive data before it forwards the plaintext input to the server, and add sensitive data on the way back out.

We built three different applications to demonstrate the applications security that this co-server application might provide.

- Jiang built a simple authentication application. A student can request his or her grade via Web form, authenticated via name and password (as responses on this form). The co-server receives the request, suppresses the password, and then forwards this re-written request to the server. The server responds with the encrypted record for that student. The co-server decrypts the record, checks the password, and if things are OK, re-encrypts the response for the client and sends it down the SSL channel.

- Jiang and fellow grad student Kazuhiro Minami extended this to an electronic voting prototype: the co-server catches the user password and uses it to authenticate the user with the *Dartmouth Name Directory (DND)* server, and also provides a way to tally votes so that the correctness of the result and the privacy of individual votes can be trusted against a malicious server operator.

- I later built a prototype of a "Box Office" server for Dartmouth. Essentially, this is a credit-card scenario that does not require changing how the credit-card acquirers accept charge information. Many entities (such as theater groups and athletics) at Dartmouth wish to sell tickets online; our computing services group is happy to set up a server, but would rather not have the liability of exposure to customer data. Consequently, they set up an Apache server that immediately PGP-encrypts customer orders and emails them to the appropriate on-campus entity. Our co-server prototype instead captures the customer data internally and (via a stripped-down and ported GPG) encrypts the data internally, so it never lives in plaintext at the server, outside the TCP.

Although somewhat successful in the lab, this co-server approach has not yet found broader use. One set of obstacles is the awkwardness of porting existing code to the CP/Q++ environment, and the difficulty of debugging both the host code and the coprocessor code in the same unified environment. A more serious set of obstacles is the fact that the internal code space is limited. For both these reasons, the modifications to Apache/mod_ssl/open_ssl were sufficiently extensive that, by the time broader deployment was considered, we would need to deeply re-examine these modifications due to upgrade creep in the software.

Furthermore, we would face the problem of the mismatch between the relatively long lifetime of a typical SSL server keypair and the relatively short lifetime of the server software configuration.

In Chapter 11, we examine a way to overcome these obstacles.

9.3   Rights Management for Big Brother's Computer
9.3.1   The Problem

Initial Motivation.   The University of Michigan's *Packet Vault* project [AUH99] sought to produce archives of LAN traffic, for later forensic analysis (e.g., in case a sysadmin needs to investigate if a newly discovered attack signature had a longer history on that LAN, or in case the university needs to respond to subpoena).

At the time, simply producing such an archive was a significant engineering feat. Beyond this, to protect the privacy of the network users, the vault encrypted each host-to-host conversation with a different symmetric key, and then encrypted all of these keys with the public key of the vault *owner*. The intent was that an adversary who obtained an archive CD-ROM could not learn plaintext contents of the traffic. The owner would make a case-by-case decision on whether contents should be revealed; if so, the owner would decrypt the appropriate content key so that the content requester (e.g., law enforcement or a sysadmin) could have access.

However, this cryptographic approach still leaves significant privacy concerns. For one thing, disclosure must occur on the granularity of a host-to-host conversation, specified by this pair of hosts. Selecting items by the plaintext contents ("packets with attack signature X ") is not possible, nor is disclosing only statistics or sanitized data. More significantly, this approach forces all stakeholders to trust in the future good will and good luck of the vault owner. A vault owner may obtain community acceptance, before starting the vault, by promising to abide by pre-agreed law and policy standards. However, should an adversary—or a rogue insider—gain access to the private key, then these promises no longer apply.

This situation is particularly troubling, given the nature of humans to try to exceed authority (and the common frequency of this in law enforcement). Indeed, the controversy over the Carnivore proposals demonstrated this tinderbox.

Broader Picture. Initially, these issues arose as some security challenges in a project some friends were working on. However, these issues also some deeper ramifications regarding socially responsible computing. In the standard DRM scenario (that engenders some much debate—recall Section 2.6), a centralized player with enormous economic powers send data to relatively powerless individuals, but tries to compel them to abide by the central player's policy when using this data. A Packet Vault is just one of many scenarios in our society when individuals share their data with a powerful central authority. Can we turn the tables, and build a DRM system that forces the central authority to abide by the individuals' rules?

9.3.2   Using a TCP

Our building block—a high-end TCP—can protect data and computation even from an adversary with direct physical access, and has hardware support for fast cryptography. These features naturally suggest a way to put armor around the packet vault.

For simplicity of presentation (and also to improve performance), we use a pair of TCPs:

- One TCP runs an *encoder* application. The Encoder uses a TCP's cryptographic abilities to quickly encrypt externally supplied data with a random symmetric key. It can use its outbound authentication features to sign (and even timestamp) this archive. The Encoder then binds the archive to an access policy and encrypts the symmetric key with public key of a designated *Decoder*.

- The second TCP runs the *Decoder* application. The Decoder accepts an encrypted archive, and a request to examine the archive. It verifies whether the archive is authentic and whether the policy bound to the archive permits this action. If so, the Decoder carries out the request within its secure boundary, prepares the result, and (if appropriate) signs it, encrypts it for the requester, and sends it back.

This TCP-based approach adds security. Because of the TCP's outbound authentication, stakeholders can have assurance that access to the sensitive data will follow the pre-agreed policy. Because of the TCP's physical security, stakeholders can have assurance that attempts by an adversary or rogue insider to otherwise gain access will result in denial (if the adversary tries the programmatic route) or the destruction of the data (if the adversary tries physical attacks instead).

This TCP-based approach also adds flexibility. Having a computational engine examine the data within a secure box gives us the ability to do searches, queries, and post-processing not easily possible with cryptographic protection alone. Even negative results—such as a signed statement saying "no records matching that pattern existed in this archive"—are now possible. Depending on how much state the Decoder can retain, we might even be able to limit queries based on time ("not more than 1 packet a second") and history ("party X can only ask three times a week").

9.3.3   Implementation Experience

Initially, I sketched out some design ideas with Charles Antonelli and Peter Honeyman of the Packet Vault project [SAH00]. However, my colleagues were suspicious of the feasibility (e.g., "I'll believe it when I see it"). This task fell to student Alex Iliev, in his senior thesis and subsequent *PET* paper [IS03b].

We built the Encoder and Decoder on an IBM 4758 Model 2, with the CP/Q++ OS layer. A Linux PC acted as the host for both. The Encoder supports operations to set up its Decoder, and to produce a TDES-encrypted archive of a network dump in libpcap format. The bulk TDES rate of the Model 2 is (in theory) sufficient to keep up with a vault on a 100Mb network.

To examine the archive, the Decoder uses Snort, which featured IP defragmentation and which examined and selected packets both using the Berkeley Packet Filter language and its own rules. We ported a subset of Snort to run inside the TCP. This task included writing simulations of POSIX syscalls that Snort required but CP/Q++ did not provide. Archive examination requests are framed as Snort rules.

In our prototype, we represent the access policy as a table; the rows represent *entry points* for access, and the columns represent parameters that a request via that entry point must satisfy. These parameters specify the required authorization, and any macro limits (such as total number of packets). Our prototype receives this policy as XML.

To simplify the prototype, we did not authenticate requests, and we did not do post-processing.

9.4 Private Information
9.4.1 The Problem

Let's reconsider the "privacy of sensitive Web activity" example from Section 9.2.2. One can think of many examples of where a server might offer a range of material to a client, but the client may regard his or her choice of material as sensitive, and want to preserve privacy of this choice.

- As above, suppose the server is offering information about health topics. In the current climate in the U.S., certain health issues (such as mental illness) are considered embarrassing in many circles, and may disqualify one from jobs and security clearances. Others (such as cancer or AIDS) can entail expensive treatment, which a thrifty but unethical employer might rather not subsidize.

  Consequently, a client might rather not have details of his health queries leaking out.

- Suppose the server is offering patent information. When an industrial research and development group is developing a new product, they often regard details of this new product as secret, until its announcement. During this period of secrecy, however, the group might engage in many related patent searches, in order to prepare their own patent applications to protect their new inventions.

  Consequently, a client might rather not have details of his patent queries leaking out.

- In many standard PKI settings, an enterprise maintains a online directory of public key certificates for its members. In applications where Alice wants to send Bob encrypted mail but has not talked to him already, she will need his certificate, and thus need to query this directory. In some scenarios, the presence or absence of a certificate in the directory itself carries meaning; for example, some enterprises revoke a certificate by removing it from the "current" directory. As a consequence, Alice might need to contact the directory for other PKI applications as well, such as verifying a signature from Bob.

If Alice or Bob would rather keep the existence of this interaction private, then they might rather not have details of these certificate queries leaking out.

In all these situations, a client might be willing to negotiate an SSL session, request their record, and receive the data through the encrypted SSL channel. The client is probably not willing to do any more work than that. However, the SSL channel only provides privacy of the plaintext against an eavesdropper (and even then, discloses that a query took place, and what the size of the record was).

What if the server were the adversary?

Given human nature, some rogue employers might happily pay a small bribe to learn which employees might soon have expensive long-term medical bills, and some server operators might happily accept such a bribe.

Given the pressures of the marketplace, some server operators might be willing to pay extra to offer a privacy-enhanced service (protected even against themselves), for the competitive advantage it might give them. Given the ideology of librarians, some libraries might happily pay extra to offer privacy-enhanced data services, in order to be able tell government agents that the library literally cannot comply with that request for an individual user's access history.

For another example:

- Internet2's *Shibboleth* middleware system seeks to enable institutions and universities) to share material over the Web in a way that conforms to the two institutions pair-wise policy, but also accommodates the legacy way that authentication at the requestor's site works. When Alice at University A wants some material at University B , she authenticates at University A the way she always does. University A then gives her an opaque *handle* X . She sends her request to University B , which then asks her university for the necessary *attributes* that X has, in order to decide whether to grant this request.

  Shibboleth uses the opaque handle in order to hide the identity of Alice from the target site, University B . However, the attribute server at her own university knows exactly who Alice is and what requests she issued—since University B told it.

  If this material is sensitive, Alice might rather not anyone know—not even the attribute server at University A .

How can a server offer such a privacy-enhanced information service? Some criteria here include:

- The client must do no more work than open an SSL session, request a record, and receive a response.

- The server must be able to provide this service for reasonably-sized dataset, and respond in reasonable time.

- The server must otherwise learn nothing about the request, except that the client made one.

(Again, in this analysis, we assume that cryptography effectively hides information.)

Naively, one might assume that simply encrypting the records would work. However, the server would still be able to learn statistics ("record 23 is the most popular") or correlations ("whoever asked for record 12 also asked for record 17"). Furthermore, if the adversary can also query the dataset, he could verify a guess about record Alice asked for by asking for one himself, and seeing which encrypted record the server fetched.

A related question is: can such a service also permit clients to add new records, and update existing ones?

9.4.2 Using a TCP: Initial View

Theoreticians have studied versions of this problem as *private information retrieval*. However, these approaches seemed too inefficient for real-world practice, and departed from the SSL model that our criteria outlined.

Dave Safford and I considered this problem [SS01] and coined the term *practical private information retrieval (PPIR)* for solutions that might satisfy these criteria. We considered using a TCP as sketched in Section 9.1:

- The protected execution and storage environments enable the TCP to negotiate the SSL session, and then to receive and handle the client's request out of sight of the host.

- The hardware to stream data across the boundary and through a symmetric cryptography engine enables the TCP to quickly fetch and output encrypted records.

- The physical security and the outbound authentication features enable the stakeholders to have assurance that the TCP is actually providing the privacy-enhanced service it alleges.

We assume that the TCP can hold only a small constant number of records in its internal storage.

We then explored how me might use a TCP to provide this service. We normalize each record to have the same length. If we assume each record is stored as a separate ciphertext on the host, and that the TCP does not retain records internally between sessions, then we're faced with an inevitable problem: to process any one query, the TCP needs to read every ciphertext. Otherwise, the server operator would know that an untouched record cannot be the one that was requested.

This analysis suggested a naive algorithm. The client tells the TCP its request over the SSL channel. The TCP reads in every record, decrypting while reading, and retains the requested one. The TCP then sends the requested record back to the client through the SSL channel.

With some thought, we can start improving this algorithm. For example, by partitioning each record into "stripes" and encrypting each separately, we can use K_1 TCP platforms in parallel to increase the speed by K_1, with a penalty for record assembly. By keeping stripes small enough that we can fit K_2 inside a TCP at once, we can handle K_2 queries at the same time, with only one pass through the data. With a little more bookkeeping, we can even accept queries at any time as we cycle through the records.

From these armchair calculations, the primary bottleneck would be the bulk symmetric crypto speed. In theory, if the IBM 4758 Model 2 could achieve its theoretical maximum of about 20MB/second TDES, then a farm of them might start making such a service feasible. Moore's Law would suggest the feasibility would increase over time.

9.4.3 Implementation Experience

I did some initial prototyping of this idea, using a 4758 with the CP/Q++ OS layer. Although it showed the idea might be feasible eventually, it also showed that feasibility was a long way off. For one thing, the crypto-across-the-boundary hardware did not appear to support quickly bringing in data through the TDES engine and *also* checking its integrity. This meant that we had to stop at regular intervals and do a separate call, from inside the device, to check a hash or MAC on the encrypted records. For another thing, the separation that CP/Q++ introduced between application code and the raw hardware made it hard to achieve the maximum hardware speed for short records, as some other work had shown [LS01].

However, this work did prompt Dmitri Asonov and Johann-Christoph Freytag at Humboldt University to start their own experiments [AF03, Asn04], leading to an ongoing series of results by Asonov and by my student Alex Iliev.

Asonov's Idea. Asonov used the TCP's private execution and storage environment to produce a permuted *shuffle* of the encrypted records.[4] That is, the server ends up holding a sequence of ciphertexts $C_1, ..., C_N$. The server knows that each C_i is the encryption of some plaintext record R_j. However, the server

---

[4]This approach is essentially the "square-root" algorithm from [GO96], as we will discuss later.

does *not* which C_i matches with R_j. Only the TCP knows the permutation, and the keys.

It is important here to stress that, in this discussion, the term *permutation* differs from the term *shuffle*.

- We use "permutation" to refer to the *function* π on the index space that captures this mapping. That is, π just takes integers to integers, such that if ciphertext C_j matches plaintext R_i, we must have $j = \pi(i)$.

- We use "shuffle" to refer to the array of ciphertexts $C_1, ..., C_N$.

Suppose the system starts out with a fresh shuffle. The TCP receives a request for record r_1 from the client. The TCP uses its private knowledge of the permutation π to determine that ciphertext $C_{\pi(r_1)}$ matches this record. The TCP retrieves this ciphertext, decrypts it internally, and checks that it is the right one and that it is not corrupted. The TCP then sends the plaintext down the SSL channel for the client. The host learns that $C_{\pi(r_1)}$ contains the record that was requested, but does not know what's in there.

When the second request r_2 comes, we have two possibilities.

- If $r_1 = r_2$, then this request is for a different record. The TCP realizes that it must look at $C_{\pi(r_2)}$. However, the TCP fetches both $C_{\pi(r_1)}$ and $C_{\pi(r_2)}$, so the host does not know whether or not this second request was the same as the first.

- If $r_1 = r_2$, then the request is for the same record, The TCP selects a random encrypted record that it has not asked for yet—say $C_{\pi(x)}$, for some $x = \{r_1, r_2\}$. The TCP and asks for both $C_{\pi(r_1)}$ and $C_{\pi(x)}$, so the host does not know whether or not this second request was the same as the first.

In general, for the kth request after a fresh shuffle, suppose $\pi(t_1), ..., \pi(t_{k-1})$ are the $k - 1$ distinct indices of the ciphertexts that the TCP has touched so far in this shuffle. To process this new request r_k, the TCP does our naive PPIR algorithm on the set consisting of these $k - 1$ ciphertexts, plus one more, $C_{\pi(t_k)}$, that appears randomly selected. Thus, we can reduce the per-query processing from $\Omega(N)$ (fetching each of N records) to $\Omega(k)$—but at the price of doing a shuffle whenever this k got too big. Asonov's scheme used an N^2 shuffling algorithm: for each C_i, the TCP reads in R_1 through R_N, and then saves and outputs $R_{\pi^{-1}(i)}$, re-encrypted.

Prototype. Alex Iliev and I decided to implement this idea, for a Dartmouth-sized directory of X.509 certificates [IS03a]. Our architecture used three IBM 4758 Model 2 platforms (although we switched from CP/Q++ to an experimental Linux kernel in Layer 2, to make development and porting easier).

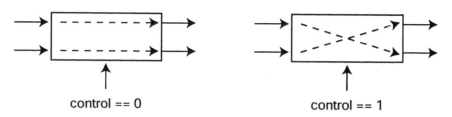

Figure 9.2. A *switch* either sends each input to its corresponding output, or switches them, depending on the setting of the control bit.

- One TCP handled the SSL connection and the LDAP requests issued by the client.

- One TCP handles each query, from a current shuffle. We also set up some hashing to handle the fact that clients tend to ask for certificates by user name, but the PPIR schemes use index numbers.

- One TCP spends its time generating new shuffles.

The sustainable query rate (or, with a bit of massaging, the maximum query response time) is determined by the time it takes to do the shuffle.

Unfortunately, the shuffling took too long. Shuffling 1000 records took 5 hours, and we estimated that shuffling 10,000 would take 3 weeks.

9.4.4   Using Oblivious Circuits

We need a quicker way for the TCP to generate an encrypted shuffle.

To do this, we started thinking about what we called *oblivious circuits*. We start with the basic idea of a *switch* gate: a box with two inputs, two outputs, and a control bit. The switch sends each input line to an output line; the setting of the control bit determines which mapping it uses. (See Figure 9.2.) We can wire these switches into acyclic circuits that take a set of input lines to a set of output lines. Any given circuit computes some family $F = \{F_S\}$ of mappings, but which element this circuit computes depends on the settings S of the switches.

Now, suppose we take such a circuit, but make each gate a black box, such that the adversary can see neither the control bit nor the internal activity of the switch. Suppose also that we make each internal wire of the circuit (as well as the outputs) encrypted tunnels, with appropriate randomness so even repeating the same data yields a different observed ciphertext. We now have an *oblivious* circuit. The adversary can observe the circuit and know that the circuit is performing some member of F. However, since the adversary does not know the switch settings S and cannot learn anything from observing the wires, he does not know which member of F it is.

Given this observation, we can take two more steps. First, our TCPs are rather good at emulating an oblivious switching network. For example:

- We put the switches in some serial order, consistent with their wiring dependence.

- We label each wire by a pair of numbers: the index of the gate it goes into and the index of the input into this gate.

- We use a good PRNG K that takes a a secret master key known only by the TCP, a nonce (also known only by the TCP) for this emulation session, and a pair of wire numbers, and then outputs a good key for our symmetric cipher. The TCP uses K to generate the key schedule for the internal wires.

- We then step through the gates. For each gate, the TCP reads in the inputs from the host, decrypts them as they come in, and checks integrity, freshness, and whether they are the right records. Based on the setting of the control bit for this switch, the TCP sends the records back out again—first output 0, then output 1, encrypted appropriately. (We need be careful about the timing to ensure that the adversary cannot use operation duration to make a good guess about the control bit.)

A circuit with G gates takes G operations to simulate. (Later, we'll use similar techniques for *oblivious sorting networks* and *oblivious merging networks*; we'll consider the problem a bit further in Section 9.6.)

For our second step, we observed that *Beneš networks* [Wak68] provide a clean way to build a compact $O(N \log N)$ gate oblivious switching circuit such that, for any permutation on N items, there exists a way to set the control bits so that the circuit carries out that permutation. (See Figure 9.3).

Thus, we can reduce our shuffling step from N^2 to $N \log N$ operations, by having the TCP generate the random permutation, calculate the control bits for this permutation, and then emulate the Beneš network obliviously. When we implemented this Beneš approach, the 3 weeks shuffle time reduced to one hour; and for the Dartmouth-sized X.509 directory, we can handle a sustained rate of one query every three seconds.

Hypothetically, we could also have used comparator gates to build an oblivious sorting network to carry out the permutation. A comparator gate decides to swap inputs by comparing some specified field within the input plaintexts (unlike a switch gate, which uses a pre-set control bit). However, the AKS family of $O(N \log n)$ sorting networks prohibitively large constant factors; Batcher's bitonic sorting networks are more reasonable to construct, but still would require a factor of $\frac{\log n}{4}$ more gates than a Beneš network.

Oblivious RAM. Back when we were discussing the encrypted-bus approaches to secure hardware (in Chapter 4), we mentioned that a natural question is how to

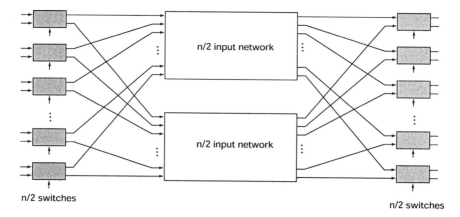

Figure 9.3. A Benes network can shuffle N inputs according to any permutation; it all depends on how we set the control bits on the O (N bgN) switches. If we make these switch gates oblivious, so that the adversary can see neither the control bits nor the plaintext on any wire after the inputs, then the network can shuffle N records obliviously.

keep the adversary from learning anything about the computation from watching the bus traffic. Ostrevsky and Goldreich [GO96, for example] developed a series of *oblivious RAM (ORAM)* algorithms to address the problem of how a CPU can an access an encrypted memory store, such that an adversary watching the bus cannot learn what the CPU is doing.

Essentially, the oblivious RAM problem is isomorphic to our way of using a TCP to carry out practical private information retrieval. The TCP corresponds to the CPU and the encrypted records correspond to the RAM; the address the CPU wishes to access corresponds to the record index the client requested via the SSL tunnel. In fact, the Asonov shuffle-and-fetch scheme corresponds to the older "square-root" ORAM algorithm of [GO96]. Our Beneš implementation provides an asymptotically and practically more efficient way of doing the shuffle; an exploration of the asymptotically superior "polylog" ORAM algorithm shows it would be inferior in practice for $N < 2^{20}$.

Asonov's Follow-on. In private communication and later in [Asn04], Asonov suggested a further improvement. If we sliced the records into small enough stripes that N could be brought into the TCP at one time, then we could use the linear-time permutation made possible by the fact that a CPU can access its addressable memory locations in constant time. We are not sure whether this approach would be effective in practice with current hardware, due to the poor performance of the symmetric crypto engines on small data sizes, and the overhead of having to knit these stripes together. However, we look forward to further experimentation.

9.4.5 Reducing TCP Memory Requirements

In more recent work [IS04b], Alex Iliev and I have pursued some different directions.

First, we considered the problem of reducing the internal memory requirements for the TCP. Let us assume we have N records of M bytes each. Already, the area of TCP-based PPIR implicit assumes that the internal memory is less than $O(NM)$: because otherwise, the TCP could just house the entire plaintext dataset. However, we also had an implicit assumption that the memory was at least $O(N \log N)$ bits.

- The TCP needs to know the random permutation, during the shuffle steps and during the retrieval steps. The natural way to write down the permutation takes $N \log N$ bits: an array indicating the destination of each entry.

- The TCP needs to be able to calculate the setting of the control bits for the Beneš network. From our exploration, it appears to be an open problem whether this can be done in less than $N \log N$ space. (We even tried instead to see if a pseudorandom setting of a Beneš network would yield a sufficiently random permutation, but without success.)

However, we developed a technique that reduces the internal TCP memory requirement to $O(k \log N)$ bits, while keeping the time requirements the same (except for a slightly longer penalty for performing the initial shuffle). Here, k is the maximum number of queries received for a shuffle. If we regard that as constant (fixed, in order to cap the maximum response time to a query), then this is arguably optimal, since the TCP requires $O(\log N)$ storage just to receive a record index.

Permutations. To start with, we cannot generate a random permutation function with only $\log N$ space. Instead, we use a short-cut: a seven round Luby-Rackoff-style cipher on $\log N$-bit blocks [LR88], using the TCP's hardware TDES engine for the pseudo-random function. With this approach, we can take a $\log N$-bit key both to a pseudo-random permutation as well as to its inverse. (In more recent work, we have been exploring the stronger properties of unbalanced LR ciphers.)

Initial Shuffle. To generate the initial shuffle, we pick a key for the permutation. Since we do not have the space to calculate the Beneš network settings for this permutation, we have to settle for a different approach. We read each plaintext record, append its destination under the permutation, and output it encrypted. We then implement a standard $O(N \log^2 N)$ sorting network obliviously. Thus, we pay $O(N \log^2 N)$ for the first shuffle, rather than than the $O(N \log N)$ we were paying before.

Subsequent Shuffles. Suppose the dataset is currently shuffled under permutation function π_1. We generate the next shuffle as follows. First, we generate a new permutation π_2. Let T be the set of records that the TCP has examined since the last shuffle. Let $\overline{\text{T}}$ be the remainder. We assume that we set a maximum constant size k for this T, in order to bound the query response time. We also assume that the TCP has internally recorded a list of the items in T (which takes $k \log N$ bits).

Intuitively, what we do might be explained by an analogy to a deck of cards. Imagine we have shuffled them, and then laid them in a row face down. Starting from the left, we have turned over the first k. This corresponds to the set T; the set $\overline{\text{T}}$ is what remains.

- In the old scheme, we would collect all the cards back into a deck, and then do an expensive re-shuffle of the whole thing.

- However, we could note that $\overline{\text{T}}$ is already pretty well-shuffled. So instead, we collect the T into a deck and do an expensive re-shuffle of that. We then collect $\overline{\text{T}}$ into a deck, and then obliviously merge the two (e.g., with a single shuffle of the two decks together).

  The adversary would know that the relative order of the $\overline{\text{T}}$ elements in the new shuffle is the same as their order before. But since he did not know what they were, it is just as random.

  This approach does not work for us, however, because we need to be able to easily calculate the mapping from index to shuffled location, and this approach of completely preserving the old $\overline{\text{T}}$ order makes it too complicated.

- So instead, we do a modification. We collect the T into a deck and do an expensive re-shuffle of that. We then just move around the $\overline{\text{T}}$ cards, face down: we pick one, and move it to the left end; we pick another, and move it after that, and so on. We pick these by the value of the plaintext index, under π_2; the point here is to get $\overline{\text{T}}$ ordered by $\pi_2(i)$.

  Then we collect these into a deck, and obliviously merge the two.

  More formally, the reshuffle consists of first processing T, then processing $\overline{\text{T}}$, and then merging.

- **Processing T.** As we noted, we assume the TCP has stored a list of the indices of the elements in T: $\pi_1(r_1), \pi_1(r_2), ..., \pi_1(r_k)$, where $r_1 .. r_k$ were the requests sent in this last round.

  The TCP reads in each ciphertext $\pi_1(r_i)$, appends it with the tag $\pi_2(r_i)$, and sends it out encrypted.

  The TCP uses this list and π_2 to calculate settings for the Beneš network that would permute these ciphertexts to the relative order these indices would have under π_2. (This takes $k \log k$ space.)

The TCP then carries out this permutation obliviously. We now have a set of ciphertexts for T, ordered by (and internally labeled with) the image of their index under π_2.

- **Processing** $\overline{T}$. From the above, the TCP builds internally a sorted list of $\pi_2(r_i)$. The TCP then walks through an index j, from 1 to N. This index corresponds to locations in the new shuffle. Because the TCP has the above sorted list, it can determine whether j is in the list or not in constant time. If not, the TCP calculates $r = \pi_1(\pi_2^{-1}(j))$, reads in the ciphertext record r (in the current shuffle), appends j as its new location, and sends it back out, re-encrypted.

We now have a set of ciphertexts for $\overline{T}$, ordered by (and internally labeled with) the image of their index under π_2.

- **Merging.** The TCP then merges these two ordered sets using an oblivious version of a merging network (e.g, as described in [CLRS01]).

The space cost of this operation is bounded by $O(k \log N)$ bits—but we assume k is constant. The time operation is bounded by $O(N \log N)$.

9.4.6 Adding the Ability to Update

We also worked on allowing the client to change record contents during a request, rather than just reading it [IS04b]. This task required dealing with three challenges:

- how to keep the server from learning which request was a write and which was read;

- how to protect against replay attacks; and

- how to deal with the fact a shuffle generated in parallel to query sessions will be stale before it can even be used.

Reads vs. Writes. We can keep the server from distinguishing reads from writes by having the TCP regard all operations as writes. When a request comes in, the TCP fetches the current "touched set" T and a new ciphertext, and re-encrypts *all* of them to produce the new "touched set" T, to be read in when handling the next request.

(This model still requires that a client not increase the record beyond some maximum size; we also do not yet let a client create a new record.)

Replay. As a consequence of allowing write operations, the ciphertext for any given record will change each time the record is touched. Naively, this situation might introduce the risk of *replay* attacks: with the multiplicity of ciphertexts for any one record, the adversary might substitute an old one—that is correctly formatted and encrypted, and will pass integrity checks, but contains data that is out of date. We considered various approaches, both simple (store big tables of hashes) and sophisticated (use Merkle trees). However, a simple and concise technique works: the TCP remembers some temporal information— which shuffle, and which query within that shuffle—then tags each plaintext with that, then checks each plaintext for the right tags.

Stale Shuffles. The biggest challenge arises when re-shuffling. In order to avoid downtime, we need to have one TCP generate the next shuffle while another handles client queries using the current shuffle. Both the re-shuffler and the query handler start with the same set of ciphertexts—but the query handler will change them during its work, making the re-shuffler's output stale.

We solve this problem by adding another dataset to the picture.

Until now, we had two datasets on the server: T and $\overline{T}$. Each record corresponds to exactly one ciphertext, that lives in one of these two sets. With each new query, the TCP moves one record from $\overline{T}$ to T, and also reads in and re-encrypts all of T in order hide the identity of the record the client was interested in. We start with a shuffle with a full $\overline{T}$ and an empty T.

Suppose a small number (k or fewer) of the records in $\overline{T}$ were incorrect, but we wanted to hide which records these were. For this shuffle, we could add a third *adjustment set* A which contains the corrected ciphertexts for these records. Now when the TCP receives a query:

- It moves a record from $\overline{T}$ to T, because the queried record could have been one of the ones in $\overline{T}$.

- It reads in all of A —because if the queried record was in $\overline{T}$, then it could have been one of the ones in $\overline{T}$ that had an error.

- It reads in all of T —because the queried record might have been one of the ones that had already been touched.

- It outputs a re-encrypted T, reflecting the update just performed.

Basically, we just have a new set A that we read in along with T each step; however, A does not change.

Suppose we started a round with a freshly shuffled dataset D_i and a correction set A_i.

- The query-handling TCP then initializes $\overline{T}_i$ to D_i, and starts handling queries based on $\overline{T}_i$ and A_i.

- The shuffle-generating TCP starts generating shuffle D_{i+1} based on D_i and A_i. There a number of ways to do this; perhaps the clearest is to pretend the client then writes each element of A_i to D_i. That is, session $i-1$ had ended with a T_{i-1} and $\overline{T_{i-1}}$. For each element of A_i, the TCP reads it in, moves an element of $\overline{T_{i-1}}$ to T_{i-1}, and then re-encrypts T_{i-1}. We end with a T'_{i-1} and $\overline{T'_{i-1}}$; we proceed with the algorithm of Section 9.4.5 to generate D_{i+1}.

- At the end of session i, we have a T_i that reflects the touches that happened to D_i corrected with A_i, and a new shuffle D_{i+1} of D_i corrected with A_i. We set A_{i+1} to T_i, and continue.

Latest Performance. We modified our prototype from Section 9.4.4 to carry out the reduced memory approach of Section 9.4.5 and the update feature of Section 9.4.6, and noticed an improvement in overall performance.

However, what we found exciting was the ability to now protect large datasets with limited memory devices—particularly given the advances of smaller TCPs (e.g. Chapter 12).

9.5   Other Projects

Other applications based on this family of TCPs have been developed and proposed. We quickly review some.

9.5.1   Postal Meters

Following in the shadow of Bennet Yee and Doug Tygar (Section 4.1.4), commercial firms including Neopostage, Pitney-Bowes, and PSI Systems have used the IBM 4758 TCP platform as a component in Internet and PC-based postage meters.

9.5.2   Kerberos KDC

Naomura Itoi used an IBM 4758 Model 1 platform with CP/Q++ to protect the *Key Distribution Center (KDC)* in the Kerberos distributed authentication protocol [Ito00]. In Kerberos, the KDC is the central trusted third party, holding all parties' keys. If the adversary can compromise the KDC, he can impersonate any party in the system.

9.5.3   Mobile Agents

Researchers including Bennet Yee have explored using a TCP to provide increased security for mobile agents [Yee99].

9.5.4   Auctions

Adrian Perrig, Dawn Song, Doug Tygar and myself examined the use of a TCP meeting the basic sketch of Section 9.1 to provide security and flexibility for electronic auctions [PSST02].

Recall the scenario of Section 2.5—each participant sends the auctioneer his or her bidding strategy, rather than casting bids in real time. The auctioneer then plays the strategies against each other and announces the winner. Why should any given participant trust that the auctioneer did not disclose his strategy to a rival? (Bob would probably be happy to know the minimum price Alice would accept for an item.) Why should the stakeholders trust that the announced result really followed from the fair competition of the submitted strategies?

For another example, consider auctions with non-trivial rules for data disclosure and winning. If the winner is to pay the second-highest bid price, how can the winner verify the auctioneer is reporting the correct value? If the auction is only supposed to reveal the name of the winner, how can the participants trust that the auctioneer will reveal no other names?

Cryptographic approaches can provide ad hoc, inefficient solutions to specific instances of some of these problems. Using a TCP can provide a general solution: the programmability gives the auctioneer flexibility to set up various types of auctions; the physical security protects the auction from external manipulation; and the outbound authentication lets the stakeholders verify that the auction might indeed follow its alleged rules.

9.5.5   Marianas

Another family of TCP application is the NSF-funded *Marianas* project, started by Dave Nicol, Chris Hawblitzel, and myself. The idea of Marianas is to merge TCPs with *peer-to-peer (P2P)* networking. P2P has been the bane of the music industry, because the self-organization and decentralization of such networks make them very hard to suppress. Using P2P between TCPs would enable us to build an overlay network that is *survivable* and *distributed* (because of P2P) but also *trustworthy* (because of the TCPs). Marianas thus might enable us take tradition trusted third party protocols and make them distributed and survivable.

As an initial exploration of this idea, John Marchesini and myself added authentication and security to Gnutella, ported this into the IBM 4758 TCP, and used the resulting P2P/TCP network to build *virtual hierarchies,* an approach to inter-CA PKI architecture that provides the resiliency of mesh architectures with the efficiency of hierarchies [MS02].

As another application of this idea, Gabe Vanrenen and myself [VS04] used P2P with TCPs to distribute Boneh and Tsudik's *Semi-Trusted Mediator* [BDTW01] approach to PKI. In this project, we built prototypes using JXTA—but did not port the prototypes into actual trusted hardware.

9.5.6   Trusted S/MIME Gateways

Another family of applications we considered was using a high-end TCP to provide a trustworthy way for an enterprise (particularly an enterprise such as a college or university, full of unmanageable free-thinkers) to add PKI to Web-based email clients.

Where should the private keys live?

- Having users carry their keys with them is expensive.

- Having users download their keys to a local client machine is awkward and, in many settings, risky. Will users delete their keys afterwards? Does the machine house a Trojan?

- Having a back-end server store and use the keys raises trust issues. Why should the user trust the server?

Furthermore, no one really wants to change their mail environment. Users of Web-based clients want to be able to check email from any machine that provides a suitable browser.

In his senior thesis, Evan Knop [Kno01] explored the idea of using a TCP-housed co-server to hold and wield private keys, for S/MIME mail. Evan also examined the logistics of trying to "multiplex" SSL sessions from a high-security co-server and a lower-security host server.

In her senior thesis, Mindy Periera [Per03] followed up on this work by exploring the use of a TCP co-server as a *gateway* between an enterprise's mail server and its clients. This approach would permit all users to continue using whatever client they wanted, and would free our PKI deployment lab from having to worry about S/MIME compatibility of all known mail clients. Mindy developed several prototypes; however, we were never able to make the footprint sufficiently small to live inside the 4758 TCPs we had available.

9.5.7   Grid Tools

The *Grid* project seeks to enable researchers at distributed universities and laboratories to share computational resources. Grid researchers have developed a suite of PKI tools to handle some of the authentication and delegation problems that result. For example, Alice at University A needs not only to "log in" somehow to a machine Bianchi at University B —she also needs to leave a long-lived process running there that can continue to act as her by accessing her data.

J. Novotony and his colleagues developed the *MyProxy* tool as a central repository for PKI credentials in Grid applications [NTW01]. To try to address the security risks of having a populations' private keys in once place, Markus Lorch and his colleagues then moved the private keys and the crypto routines that handled them into an IBM 4758 [LBK04]. More recently, John Marchesini here at Dartmouth has been working on moving the computation that calls these routines into the TCP, as well as extending to other types of TCPs [Mar04].

9.6   Lessons Learned

We conclude this chapter with a few observations gained from this experience designing and building applications.

Programming Environment.   The closer the TCP could provide to a standard programming environment, the easier the process was—and the easier it was to build and test ideas, rather than having them languish on the shelf. For one example, switching from CP/Q++ to Linux for our 4758 experiments let us use the standard GNU suite of build tools and debuggers, and made it far easier to port existing software into our environment.

Threat and Failure Models.   When considering whether and how to deploy these applications in real settings, questions emerged regarding what may happen should these TCPs fail. In the case Itoi's hardened KDC, one colleague recalled a model of disk drive where *every unit* failed, due to a component problem, and wondered: what if every IBM 4758 fails, due to some hardware problem? Even if the design works and these units treat the failure as a tamper event, an enterprise that used such a TCP as the sole holder of the KDC could have a problem. This scenario suggests the need for some thinking about how to use heterogeneous TCPs to increase reliability without decreasing security.

In our Marianas project, which targets eventually using heterogeneous types of TCPs, the question arose about what happens should an adversary compromise a platform. In the current information infrastructure, security flaws usually do not affect just a single instance of a system, but rather compromise a large family of systems—any instance with that flawed software, for example. In our current estimation of physical security attacks on TCPs, we speculate that the adversary might need to take such a TCP into a laboratory and experiment a bit before discovering a successful attack. Together, these scenarios suggest the need for some thinking about how to model the threats for such a system: for example, perhaps we should assume that if a particular device disappears for some period of time, then the probability increases that all devices of that class may be compromised.

New TCP Models.   One of the exciting aspects of our private information work (Section 9.4) was its suggestion of the value of new computational models.

On a basic level, these applications showed that theoretical security techniques (such as oblivious RAM) that had been deemed too impractical to ever find real-world use, could indeed find real-world use. What other theoretical techniques can we mine? For example, theoretical PIR algorithms exist which, at first glance, seem to require too many parties or too much work by the client—but which might become quite feasible if we insert a TCP as a proxy in the right place.

On another level, the development of this idea of *oblivious circuits*—and these security-relevant applications of dusty old ideas such as Beneš networks and merge networks—suggests that a richer application space may lie here. The usefulness of oblivious circuits also suggests the potential for developing a TCP customized just for this purpose:

- limiting the memory and CPU, to save cost (and since our work showed that we might not need that much memory anyway)

- but increasing the ability to quickly bring data in and out across the secure boundary, through a symmetric crypto engine that also checks for integrity, and perhaps for some other formatting flags.

9.7 Further Reading

The litany of WebALPS applications in Section 9.2.2 was based on Section 3 in my *SIGEcom* paper [Smi01]. Shan Jiang's thesis [Jia01] and our conference paper [JSM01] provide more discussion about the design and prototype. Eric Rescorla's book [Res00] is an excellent reference for SSL.

More details on the Armored Data Vault work appear in our *PET2002* paper [IS03b]. Section 9.4 cites the principal TCP-based private server approaches: Asnonov's [AF03, Asn04] and ours [SS01, IS03a, IS04b]. Our *IEEE Security and Privacy* paper surveys both the armored vault and the privacy server work [IS04a].

Chapter 10

TCPA/TCG

This book has focused on trustworthy computing platforms. What does it mean for stakeholders to be able to trust a computing device to carry out its correct computation, despite direct adversarial attack? How can we go about building such a device? If we had one, what could we do with it?

This book has also emphasized real experience. We do not want to just reason about what might be possible; we want to try to make this real, in the real world. This drive motivated the development of the 4758 platform (Chapter 5, Chapter 6, Chapter 7), establishing assurance (Chapter 8) and exploring applications (Chapter 9).

This experience left us with several observations:

- A TCP can indeed enable practical solutions to distributed security problems.

- We prototyped many of these with the IBM 4758 platform. However, in theory, any platform providing the basic core functionality of a TCP should be able to support these applications.

- The limited-power, special-purpose programming environment of the 4758 hindered its usefulness. Programming and debugging this embedded system via the "soda straw" the CP/Q++ developer's kit offered is awkward. Porting existing software to an OS that does not provide standard syscalls is awkward. Even with the Linux, the limited internal code-space hampered development of our S/MIME gateway prototypes.

- The relatively high cost of a 4758 (e.g., a Model 2 device retailed for about US$3K, in quantities of one) and relatively large size (e.g., this PCI card will never fit in a laptop or PDA) will keep it from ever being ubiquitous.

However, recent years have brought the gradual emergence of another family of device that can potentially be the foundation of another family of TCPs that address these concerns, at the cost of a lower threshold of security.

In 1999, the *Trusted Computing Platform Alliance (TCPA)*, a consortium of many leading hardware and software vendors, formed and began considering the problem of how to increase the level of security on commodity desktop and laptop machines, without significantly affecting the cost. They developed an architecture based on a *Trusted Platform Module (TPM)*, currently existing as a small chip added to the motherboard that assists in witnessing and perhaps controlling the boot process (in the spirit of the early work by Yee, Arbaugh, and others Section 4.3), binding secrets to machine configuration and attesting about this configuration to other parties (in the spirit of our outbound authentication work Chapter 7). In 2003, the *Trusted Computing Group (TCG)* took up the mission. However, in the popular lingo, the term "TCPA"—the *old* name of the *organization*—persists as the name for the design and vision.

The TCPA/TCG approach is exciting because it's real and because (for the most part) it's open. TPMs exist; many commodity machines ship with them already installed; the organization has published the specifications for both the TPM as well as the *TCG Software Stack (TSS)* to be supported by the TPM. On the other hand, the TCPA/TCG approach is challenging partly because of the large number of players involved, and because of its potential role in a large family of emerging products. A tempting hypothesis is that the multiplicity of authors—and the lack of one driving clear vision—makes the specifications hard to read and digest. Another tempting hypothesis is that the TCPA/TCG design supports a product vision that has not been fully articulated yet publicly (nor, perhaps, privately), and the pieces do not quite hold together for those of us outside lacking that keystone. As another artifact of this industrial reality, the TCPA/TCG vision has been a moving target. The TPM has gone through several specifications. The currently available hardware does not match the latest specification (and, sometimes, it does not quite match its own specification).

Nonetheless, the TCPA/TCG approach—put a small, cheap chip on the motherboard and use it to secure a commodity desktop—targets a different point in our TCP solution space. This chapter presents a summary snapshot of the TCPA/TCG approach; Chapter 11 will discuss our experiments in trying to turn current TCPA/TCG hardware into a "virtual" 4758 and use this TCP for applications.

Section 10.1 discusses the basic structure of the TCPA/TCG architecture. Section 10.2 discusses how we can use the attestation features of TCPA/TCG to provide outbound authentication functionality. Section 10.3 discusses the physical security profile of this architecture. Section 10.5 gives some background on starting experimentation. Section 10.6 discusses some of the interesting changes promised in the new 1.2 spec.

Caveat. It is important to note that, by definition, the material in this chapter will be obsolete. As noted above, the TCPA/TCG design is a continually moving target. The interested reader is urged to consult the Trusted Computing Group for the latest specifications and thinking.

10.1 Basic Structure

We'll start by reviewing the TCPA/TCG hardware design, based on the 1.1b hardware specification [Tru02] for the PC platform [Tru01]—since this has been what's been available for experimentation. To increase clarity for this discussion, we'll avoid the generality a specification can introduce, and instead take a practical systems perspective: this is what you see, if (like us) you buy a TPM-equipped PC and start playing.

The Trusted Platform Module. As we have noted, the heart of the TCPA/TCG design is the trusted platform module. In current instantiations, the TPM is a smart-card-like chip that lives on the motherboard of a general-purpose desktop machine. As with our hardware lock microcontroller in the IBM 4758, this architecture assumes that the TPM knows when the system is rebooted. The TPM (at least the ones we have been working with) possesses the ability to perform SHA-1 hashing and some (slow) RSA operations; it also has a hardware-based RNG.

Like a 4758 layer, the TPM goes through a process through which it receives *ownership*. The owner possesses exclusive privilege to request certain TPM operations; the owner authenticates these requests by using a secret 20-byte key to generate a keyed MAC value on them. (The TPM uses HMAC based on SHA-1). This process also results in the creation of a *storage root key (SRK)* for the TPM.

Once owned, the TPM's primary functionality centers on its ability to witness platform configuration and to provide protected storage of keys and data.

As with 4758 layers, the TPM "owner" is the party in charge of administrating the security of that TPM—and in particular is not necessarily the user of the machine.

Platform Configuration. The TPM's vehicle for witnessing platform configuration is a suite of *platform configuration registers (PCRs)*. Each PCR is 20 bytes long, the length of a SHA-1 hash value. At boot time, the TPM resets the PCRs to zeros. After that, the main CPU can then "write" new values to the PCRs. However, rather than replacing the current value, a "write" of a new value *extends* the old value. That is: suppose the PCR in question currently contains value v_0, and the CPU wants to update it with v_1. The TPM concatenates v_0 with v_1, calculates the SHA-1 hash v_2 of that concatenation, and stores v_2 in the PCR. This new PCR value v_2 thus embodies v_0 then modified by v_1, but in a way that (should the intractability assumptions underlying hash functions be correct) the adversary would not be able to calculate a v_1 that would take him a desired v_2.

This hash-and-extend operation thus greatly generalizes the ratchet lock mechanism in the 4758. Not only can the CPU advance a PCR in a way that cannot be reversed; the CPU can also advance it in different ways, depending on the v_2 used, and the CPU has a choice of PCRs with which to work. Because this latter feature, the PCR suite can represent not just the current state of the system within some execution progression; it can also represent choices in the path to that point. For example, in the 4758 ratchet lock, the CPU advances the ratchet from 0 to 1 before moving from the ROM phase of Miniboot to the FLASH phase, and from 1 to 2 before advancing to the OS. With a PCR approach, the CPU could also update a PCR twice: but the updates could now also reflect the *identity* of the next block of code, as represented by its hash.

In the TCPA/TCG approach, the platform uses the PCR suite to record the execution sequence and the software and potentially hardware involved.

The platform does this inductively. Initially, the BIOS reports its hash to the TPM—thus the TPM's PCR suite reflects the configuration of this system, running BIOS. The BIOS sends the next block of code to the TPM, to hash and then incorporate into the PCRs—thus the TPM's PCR suite now reflects that subsequent configuration, unless the first one was corrupt. The process continues. If the BIOS is adversarial, it can thus subvert this entire process. Hence, in TCPA/TCG lingo, the BIOS is the *root of trust for measurement (RTM)*.

The basic specifications leave the designer some latitude in deciding what to measure and how to incorporate these measurements in the PCR suite. Furthermore, as noted above, the PCR suite can also include hardware measurements; we noticed this empirically when we changed a memory card in our test machine and saw the PCR values change.

The platform can thus use the PCR value to speak about its configuration. In TCPA/TCG lingo, the platform supports *authenticated boot* when it can prove to a third party what software booted; the platform supports that apparently synonymous *secure boot* when the TPM ensures that the system does not boot unless the correct PCR suite results. (We were not able to determine how the TPM in our machine might do this.)

Protected Storage. Using these PCRs, the TPM provides the designer with a way to bind stored data to the platform only when it is in a specific configuration. At a high level, one can think of this a general *sealing* and *unsealing* service. The platform can *seal* data by listing which PCRs are of interest and providing an additional 20-byte authorization code. The TPM responds with an encrypted item. To decrypt this item, the platform needs to present it to the TPM—except the designated PCRs must have the same values they had when the data was sealed, and the caller must know the authorization code. (At a lower level, the TPM provides this service via an internal key hierarchy rooted in the SRK; it can be easy for a reader to get lost in the command set that deals with these keys.)

Since the TPM has the ability to do RSA operations, it only makes sense to share that with the main platform. Some services of interest include:

- The TPM can create a key pair and seal it to a given configuration, as one operation.

- If the TPM knows a protected item is a private key, it can do a private key operation with it instead of revealing it, when the caller is authorized (that is, when the PCRs are correct and the caller knows the authorization code).

- If the caller is authorized, the TPM can use a protected private key to sign a statement attesting to the current PCR values.

- If the caller is authorized, the TPM can use a protected private key signing a certificate about the public part of a TPM key pair—including the PCR policy for its private part.

Credentials. The platform can use its TPM to build up a PCR suite that reflects its current configuration, and can bind keys and seal data items to such representations of this configuration.

However, to convince a remote relying party, the platform needs to somehow bind these signed assertions and hash values to something the remote relying party trusts. The TCPA/TCG architecture achieves this through a set of *credentials*.

- The TPM leaves the factory with an *endorsement key pair* whose private key (one hopes) is uniquely known by that TPM.

- The *endorsement credential* (typically from the TPM manufacturer) is a signed statement by some trustworthy entity that binds the endorsement public key to that type of TPM.

- The *conformance credential* states that the TPM and its platform conforms to the specification (and can be "issued by anyone with sufficient credibility" to do this evaluation [Tru04]).

- The *platform credential* (typically from the platform manufacturer) contains references to the endorsement credential and the conformance credential, some general statements about the platform type.

■ An *identity credential* is a signed statement from what the TCPA/TCG spec calls the *Privacy CA* that binds data about the TPM to a special identity public key. (We'll discuss this shortly.)

Although the credential set appears complex, a view of the industrial process (and the backlash Intel faced for earlier trying to include machine-readable serial numbers in its CPUs) might appear to simplify things. First, one might want to separate the TPM manufacturing and blessing process from the machine manufacturing process, hence the separation of endorsement credential from platform credential. Second, one might want to separate the end application use from these manufacturers, in order to preserve the privacy of the deployer and the users; hence the separation of the platform credential from the identity credential, and the use of the term "Privacy" in "Privacy CA." (In our experimental work, we used the alternate term *Yet Another CA—YACA*.)

Although the specification and documentation discuss this large suite, the only thing that the TPM in the machines we experimented with appeared to come with was the endorsement key pair. From our reading and testing, it appears that the TPM will only use the endorsement private key for decryption (not digital signatures), and only in the context of the TPM_TakeOwnership and TPM_ActivateIdentity commands.

10.2   Outbound Authentication

As we observed back in Chapter 7, perhaps the simplest way for a TCP application entity to prove that it is "the real thing doing the right thing" is to wield a private key that exclusively belongs to it (thanks to the platform security architecture) and whose public key a relying party believes belongs to it (thanks to some trusted CA, and perhaps the platform security architecture as well).

In the TCPA/TCG architecture, *identity key pairs* play this role. The TPM Owner establishes such a key pair and its binding through a multi-step process.

First, the TPM_MakeIdentity command causes the TPM to generate a new key pair and an *identity binding* for it. The identity binding contains:

■ the newly generated public key;

■ a name for this entity, chosen by the caller; and

■ the public key of the CA that will certify this identity.

All of this is signed by the new private key.

The TPM Owner then packs this data along with the credentials and ships it off the CA. The CA inspects everything to see if it all makes sense. However, at this point, the CA cannot really conclude that the claimed key pair belongs to the platform that the credentials describe with the name the owner claimed. It might—but then again, maybe this is all a clever fraud.

Consequently, the CA hedges its bets. It signs an *attestation credential* testifying to this new identity. However, the CA encrypts this credential with a random symmetric key, and then encrypts this symmetric key with the endorsement public key of the TPM allegedly involved. (Remember, the endorsement credential provided a signed assertion from the TPM god saying that a genuine TPM had this public key.)

At this point, a signed certificate exists for this entity—an "identity" allegedly living on a particular platform with a particular TPM. However, only that TPM is in a position decrypt the symmetric key that encrypts this certificate and release it to the outside world. The TPM Owner handles this action with the TPM_ActivateIdentity command.

At this point, the platform can ask the TPM to use the identity key to sign special types of assertions. (The TPM will *not* use it to sign arbitrary data.)

We can use this to reproduce the outbound authentication functionality of the 4758 TCP: an "OA Manager" obtains this key pair, then (upon request of other modules) creates new TPM-held key pairs, wrapped to PCR value suites reflecting the calling module on this platform, and uses the identity private key to certify (via TPM_CertifyKey) these new public keys and their bindings.

10.3   Physical Attacks

Clearly, as a single chip sitting on a motherboard within a standard unprotected commodity machine, the currently available TPMs do not provide the same level of protection against an adversary that higher-end TCP predecessors do.

Several avenues of attack immediately suggest themselves.

- With the possible exception of TPM-held private keys, computation is still occurring outside the physical boundary of the TPM. Tools such as logic analyzers can still reveal secrets.

- The TPM suffers from a classic TOCTOU risk: what if the code changes between the time the TPM records the hash of a block of code in a PCR, and the time that code uses its TPM privileges? Malicious DMA and dual-ported RAM both might be effective attack tools here.

- A Chapter 3 discussed, trusted hardware modules—particular single-chip devices—have a long history of susceptibility to side-channel and induced-fault attacks. How resilient is any given TPM?

On the other hand, one gets what one pays for. Rather than asserting that TCPA/TCG provides no protection against hardware attacks (as some colleagues have claimed), I would rather think of this design as just a different tradeoff in the cost versus power versus security space. The barrier is higher than with a standard machine, Furthermore, the potential exists that a future-generation TPM, perhaps in conspiracy with security-enhanced CPUs (Chapter 12), might raise the bar even further.

10.4   Applications

It is tempting to believe that the TCPA/TCG architecture was conceived specifically to combat software and music piracy: operating system installations and media applications can set themselves up with TPM-protected and TPM-certified key pairs, and then use these to carry out the sorts of protected exchanges of software and data files envisioned several decades ago (Chapter 4). Indeed, as Chapter 2 reviewed, many in the field oppose this technology explicitly because of the vision of a dark future where consumers and end users have little control over their information vehicles.

Alternatively, rather than being cynical, one could also be visionary, and speculate on the similarities of a TCPA/TCG platform to the earlier high-end platforms. Could we turn a TCPA/TCG desktop into a cheap, powerful, ubiquitous (and insecure) 4758, and proceed to build applications like those in Chapter 9? Chapter 11 will consider this vein.

Meanwhile, TCPA/TCG architects have been suggesting applications of their own. For one example, an enterprise may require a client to attest to its software configuration before letting it join the enterprise network, in order to prevent unpatched and potentially infected machines from compromising network security.

Chapter 2 in [Pea03] provides a broader survey.

10.5   Experimentation

As the beginning of this chapter noted, a large appeal of the TCPA/TCG architecture—even the 1.1b TPM, with no official OS support nor applications yet—is that it exists in commodity hardware, and one can experiment with it.

A would-be experimenter needs to determine whether a particular platform actually holds a TPM, and to which specification it adheres. IBM released an open-source Linux driver for the TPM [IBM], which we used. In our lab, student Rich MacDonald started with the specifications and began producing an open-source library to enable platform code to format TPM calls and process responses. Two years later, just as he was finishing, IBM released an open-source library [SKv03].

Some obstacles we faced included the platform originally shipped with the wrong BIOS for the TPM; the TPM's endianness does not match the x86; mistakes in command preparation triggered rather unhelpful error codes; and some of the TPM commands generated "call not implemented" errors.

10.6 TPM 1.2 Changes

As noted, the TCPA/TPM architecture is a moving target. As of this writing, the 1.2 specifications for the TPM have been published [Tru03a, Tru03b, Tru03c], although TPMs compliant with this new specification have not been available for experimentation.

The new specification offers several new features that could prove interesting for building TCP architectures. We quickly review some.

Locality. One of the difficulties in the 1.1b design (and also in the 4758's ratchet locks) is the inability to switch back and forth between two trust contexts. If we advance the ratchet or extend a PCR to put away context A's secrets when we switch to context B, we cannot go back to A later. The 1.2 specification includes a notion of *locality* that (apparently with some hardware support) would permit the CPU to have the TPM switch back and forth between contexts. The new specification also includes a TPM_PCR_Reset command, which requires special locality privilege; however, apparently only some the PCRs may be reset.

Monotonic Counters. Another difficulty we had with building on the 1.1b design was defending against various types of replay attacks, and ensuring that our system's trusted core can know that state is advancing, and certain data or assertions may be stale. The new specification includes *monotonic counters* that might be useful here. (Interestingly, we also included monotonic counters in the 4758—but that was because application constraints forced us to expose control of the real-time clock to possibly adversarial Layer 3 code.)

Delegation. Another difficulty we had building on the 1.1b design was the fact that we sometimes wanted several different entities requesting commands restricted to the TPM Owner; this would appear to require sharing the TPM Owner's 20-byte HMAC key. The 1.2 specification includes a detailed *delegation model* that might help with this problem.

10.7 Further Reading

[Fel03, Sch02] provides short, balanced overviews of the TCPA/TCG architecture. The Hewlett-Packard book [Pea03] provides a longer discussion. The specifications are more definitive, but not quite as readable. (One student commented that the specification says "End of Informative Comment" but nonetheless goes on for hundreds of pages. The recent "Architecture Overview" document [Tru04] appears to be much-needed.)

The preliminary technical report [MSWM03] on our experimentation also provides a narrative about TCPA; Section 10.1 and Section 10.2 above were loosely based on Section 3 in that report.

Chapter 11

EXPERIMENTING WITH TCPA/TCG

Their potential to solve real security problems in the real world drove much of the research and commercial development in trusted computing platforms. Chapter 9 discussed our attempts to use a high-end TCP for some of these applications. That chapter concluded by observing that the limitations of a small, expensive device hindered application development and deployment. Chapter 10 then gave an overview of the TCPA/TCG architecture, and its inexpensive TPM hardware that is commercially available and can potentially turn an entire commodity machine into a TCP (albeit of lesser physical security).

The questions naturally arise: how real is this potential? Can we use this hardware to build a generic TCP, and build applications that the limitations of a 4758 made impractical? This chapter will discuss my group's experiments to do just that: using the commercially ubiquitous 1.1b TPM [Tru02, Tru01] to secure a larger Linux desktop.

- First, we consider the general challenges: Section 11.1 frames the problem, and Section 11.2 considers the lifetime mismatch that arises when we try use TCPs for complex applications.

- Then, we discuss the platform we built: Section 11.3 presents the architecture we developed, and Section 11.4 presents our implementation experience.

- Finally, we discuss applications we built on these platforms: Section 11.5 reconsiders the hardened Apache application of Section 9.2, Section 11.6 discusses using this platform to harden OpenCA, and Section 11.7 discusses using it to balance DRM needs with user privacy needs.

11.1   Desired Properties

First, we need to consider what properties we want from our desktop TCP.

Computational Power. We want the convenience of being able to run an OS sufficiently similar to common commodity systems to enable easy adoption of legacy software and applications.

Trustworthy Computing Environment. The platform needs to be as trustworthy as possible, within the basic limits that the machine is still susceptible to "logic analyzer" attacks (Section 10.3).

Outbound Authentication. An application entity running on our desktop TCP needs to be able to prove it's "the real thing, doing the right thing."

As we have discussed in Chapter 7, it is probably cleanest if an application entity can do this by possessing exclusive use of a private key, whose public key is supported by some certificate chain that establishes the platform and configuration parameters necessary to convince the relying party to trust the entity for this application.

Secure Storage. The desktop platform also needs to be able to store non-volatile data, and either zeroize it or otherwise render it unavailable to the adversary upon attack. (Of course, we can only expect the platform to handle attacks within the threat model in a TPM architecture.)

Deployment. As with the 4758, we want to enable guerilla-style development and deployment of applications. That is, we do not want to require the hardware vendors and major software vendors to be committed to major involvement in any given application.

Supporting experimentation and development was one of the reasons we chose open source, and decided not to wait for the TSS stack to be released — particularly since the specification itself had not been made public when we started. Too much debate focused on the negative potential of this technology (e.g., Section 2.6), even though TCPs had a considerable positive potential (Chapter 2, Chapter 9). We hoped to enable experimentation to supplant idealogical debate (and also appreciated the potential irony of open source use of TCPA/TCG hardware to take root before commercial closed-source systems catch up).

11.2   The Lifetime Mismatch

Relying parties like to make trust judgments about entities that know the private key matching a public key with some type of trust chain supporting it. To simplify the process for the relying party, this trust chain often reduces to a certificate from some CA the client has chosen to trust for this purpose. Because of its importance, this certification is not a lightweight process. Consider an SSL-protected Web server: to certify the public key, the sysadmin must go through a lengthy process with a commercial CA, and then find someone with room in their budget for the CA's fee.

Our general TCP model adopts this scheme: an application somehow ends up with such a private key, and the TCP architecture confines that private key to that application. However, we want to harness an entire desktop machine for our TCP, and we want to provide something close to a standard programming environment for its applications. The intended consequence of such an environment is to make it easier to end up with standard programming: large applications and code libraries, maybe even a large operating system. As enterprise security officers and conscientious sysadmins know all too well, a seemingly inevitable property of our current complex software environments is an endless stream of vulnerabilities. For an administrator of an application on a TCP that permits such complex software, the responsible action is to vigilantly monitor announcements and apply patches, if this application entity is to be trustworthy.

The action necessary to maintain a trustworthy environment conflicts with the way the TCPA/TCG platform confines secret data to application entities—via hashes of the software itself. Consider again the SSL Web server example. A responsible server installation may go through multiple upgrades of the software stack in a year, but keep the same key pair for many years. How do we move from secrets bound to short-lived software configurations to a private key bound to some long-lived "trustworthy" entity? How does the relying party draw this conclusion from the certification attached to the long-lived key? How does the CA, when it certifies the key, trust that the sysadmin will continue to be conscientious? When a once-trusted system configuration becomes untrusted, due to a vulnerability announcement and patch, how does the TCP prevent this insecure version from retaining the key—or a malicious operator from rolling back a patched configuration to the vulnerable one, in order to exploit the vulnerability?

Besides the basic code base, we also need to worry about the dynamic nature of other data and parameters that the application depends on. Consider the SSL Web server again—even if the TCP protects the Apache and SSL code, what about the pages and CGI scripts that the server offers?

11.3 Architecture

Our architecture thus needs to build on the 1.1b TPM to provide three things:

- a trustworthy environment, based on a reasonably full, reasonably standard OS;

- a way to bridge the above mismatch between entity lifetime and configuration lifetime; and

- a way to provide secure storage, bound to the entity (not the specific configuration).

Trustworthy Environment. We begin by considering the trustworthy environment. As noted earlier, we started with Linux, as perhaps the most dominant open-source OS; the dominance means applications and software exists, and the open-source status means that we can get easily dive in and modify internals.

To ensure the system with a complex software configuration remains in a safe state, we drew on previous work in kernel integrity (e.g., [BBC[+] 00, vGA01]): an *Enforcer* module checks every file and directory against what we call a *configuration file*: a signed table of hashes. To try to minimize the performance hit, we perform these checks when a file is first opened, by hooking inode permissions checks. To ensure the Enforcer itself runs correctly, we use the TPM PCRs to record the execution sequence up to and including the Enforcer and the kernel code it depends on, and we build the Enforcer into the kernel itself, within the *Linux Security Module (LSM)* reference monitor framework. (This LSM framework also allows us to extend the Enforcer to hook and perform checks on other system calls as well.) To ensure that Linux boots securely, we compile it into the kernel (rather than using loadable modules); as Section 11.4 below discusses, we also need to make sure that the adversary cannot subvert the process by changing boot parameters.

We named this Linux-based TPM platform *Bear*; the overall project came to be known as *Bear/Enforcer*.

Entity Mismatch. To address the lifetime mismatch, we use the TPM PCR suite to protect the stable core: the kernel and the Enforcer module. The Enforcer verifies the authenticity of the configuration file via a digital signature. A *Security Admin*—who may very well be a remote third party, such as Thawte or one of the other commercial CAs—owns the private key, and produces these signed configurations of what it deems to be a secure configuration. For now, we hash the public key into a PCR value, as part of this stable core.

The CA that signs this platform's public key is testifying both to this platform and long-lived core, as well as to the continuing good judgment of the Security Admin.

We plan also to incorporate a *low-water mark*. Each signed configuration file would include a current serial number, as well as a new low-water mark. After verifying the signature, the Enforcer replaces its TPM's current low-water mark with the pairwise maximum of that value and the one in the file, and then only accepts the file if its serial number not less than that minimum. Our intention was to allow a sysadmin to roll back a new update when appropriate (e.g., due to incompatibility of the update with needed functionality), but still to enable the Security Admin to establish a minimum baseline. It would be interesting to explore using additional techniques (e.g., perhaps based on the monotonic counters in the 1.2 specification) to make this revocation of old configurations more automatic. In some sense, this approach is just a generalization of the "epoch" entity approach we developed in the 4758 (Chapter 7), just as the TPM PCR approach generalizes the "configuration" entity approach (Section 10.1).

Implementing our low-water mark approach will require solving the problem of TPM-controlled non-volatile storage. In 1.1b, we explored the use of the *Data Integrity Registers*, which had murky functionality which have since been deprecated.

Secure Storage. The platform also needs to provide secure storage to its on-board applications, in a way that is easy for the application programmer to exploit. In modern operating systems, the standard paradigm for non-volatile storage is the filesystem. The tool of a *loopback filesystem* lets us provide a filesystem to the application programmer, while appearing as a file to the machine's OS; an *encrypted loopback filesystem* keeps this file encrypted. To provide secure storage, we used an encrypted loopback filesystem. The TPM protects the symmetric key that encrypts the loopback. (In our prototype, we used AES.) The Enforcer only releases this key if the configuration checks are successful. (We are exploring various ways of ensuring the freshness of this encrypted file.)

Putting it All Together. To put it all together, we looked at the components of the platform and partitioned them by lifetime.

- **Long-lived data.** The initial BIOS ROM (the root of trust management) reports itself to the TPM. The TPM then inductively measures the rest of the boot sequence up to and including the Enforcer.

- **Medium-lived data.** The Enforcer measures the public key of the Security Admin by hashing it into a PCR. The Enforcer then checks the rest of the rest of the security-critical software against the configuration file. If things are OK, the current PCR suite enables release (or TPM use) of the application's private key and other secrets; if things are not OK, the Enforcer advances the hash and (if we had released the application secrets into the system memory) zeroizes secrets.

- **Short-lived data.** If the PCR values are correct, the TPM will release the symmetric key for the encrypted loopback filesystem. The Enforcer mounts the filesystem and makes it available for use by the application. The OS may allow the application and other users to read and modify data in the

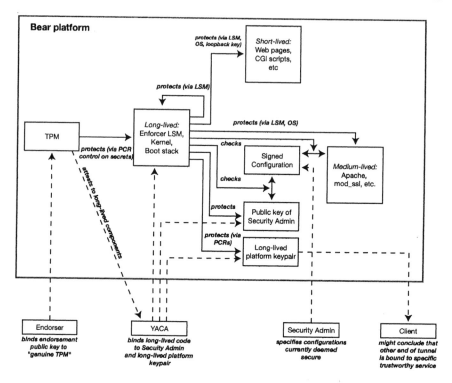

Figure 11.1. Sketch of the flow of protection and trust in our platform. To enable a client to make a trust decision about dynamic content based on a long-lived application key pair, we introduce indirection between the core long-lived components and the more dynamic components. Our intention is, like the IBM4758, the TPM/Linux platform would let the end user (here, the server operator) to buy the hardware and install the software; our platform could authenticate these components to "Yet Another CA." Here the dashed lines represent interactions between the platform and external parties.

encrypted loopback filesystem; however, the Enforcer has checked these item.

To provide outbound authentication, we can follow the process that Section 10.2 outlined: we set up a TPM identity key pair wrapped to the long-lived core, have an external CA sign a certificate testify to this, and build from there.

Figure 11.1 sketches the overall structure.

In our latest experimental version, we also allow the configuration to specify that "only application X can access file Y " (e.g., the "poor man's compartment").

11.4 Implementation Experience

We built the Enforcer as an LSM for the 2.6 kernel (or a 2.4 kernel with the LSM 2.4.20-1 kernel patch). The initial prototype is about 2000 lines of code. As noted above, for secure deployment, the Enforcer should be compiled into the kernel; however, we also included hooks to load it is a separate module, in order to simplify development and testing.

As Section 10.1 discusses, at boot time, the BIOS reports itself and the *master boot record (MBR)* to the TPM, and then passes control to the MBR.

We started with the LILO loader. We modified the first-stage bootloader first.b (the MBR in LILO) to hash second.b. Doing this in assembly, to fit within the tight confines of the MBR and handle the TPM Endianness requirements, was tricky. In our current prototype, we run our TCPA-enabled LILO from a floppy. This decision stemmed in part from codespace—and also because of problems with IBM's "Memory Present Driver," which we used to communicate with the TPM via real-mode assembly. An MBR is 512 bytes; but a hard disk MBR also contains other data (the disk partition table or tables), and does not give us the full 512 bytes for code. This would not have been a problem, except the TPM in our machine did not appear to actually support the TCPA_HashLogExtendEvent call; we kept getting a "call not implemented" error. The workaround—replacing this call with a sequence of calls—pushed us over the limit for the hard disk MBR. (The floppy gives us the full 512 bytes.)

We modified second.b to passing the memory where the kernel resides to the TPM, so that the TPM hashes the contents of that memory into a PCR. This hash will cover any persistent data the loader uses (i.e., all of the information in /etc/lilo.conf). The LILO program (that is, /sbin/lilo) reads the data in the configuration file and merges the data with the actual loader (first.b and second.b)—because no filesystem exists at the time the loader is booting the system so the necessary information is put into the data segment of the loader program itself.

If the adversary wants maliciously change the boot process by changing lilo.conf, then /sbin/lilo must be run to make those changes take effect, and the system must be rebooted. However, if all this happens, the PCR which holds the kernel's hash will hold a different hash value than the previous boot.

The adversary could also try to pass malicious arguments to the loader via the command line at the boot prompt, but second.b can include these arguments in the kernel hash, and thus embed them in the TPM's perception of the platform configuration. Alternatively, the adversary could try to change the mapfile. However, our system calculates the hash of the kernel image once it is in memory, so such an attack will place a different hash value in the PCR. Again, our system will not allow the secret to leave the TPM.

The Enforcer uses the /etc/enforcer/ directory to store its signed configuration file, public key, etc. Having the kernel store data in the filesystem is a bit uncouth, but seemed the best solution for this problem—and also is not completely unprecedented.

In addition the platform itself, we also wrote a number of small executables which make some of the TCPA calls necessary for attestation (TPM_MakeIdentity and TPM_ActivateIdentity). We also wrote some utilities to produce the security policies, and for each file covered by the policy, the Security Admin can specify what should happen if its integrity check fails: log, deny, or panic.

We developed the Enforcer under user-mode Linux (UML), which worked very nicely—each bug that appeared under UML also showed up with the real system, and vice-versa. We ran basic functional tests, showing that modifying the configuration file, public key, signature, or any protected file actually causes the appropriate reaction. We also ran 36 hours of continuous stress-tests; the code showed no signs of crashing or leaking memory.

Issues. At the time of this writing, several areas remain for future work in our design and implementation. We might improve performance by clever caching of validated hashes. We also need to examine more carefully how to use this signed-configuration approach to handle components—such as logs—whose contents will change during correct execution. Right now, the best solution appears to be to put these into the short-lived loopback—and work on clever ways to ensure freshness of the data in this loopback file.

The major limitation of our approach is that data elements that the Enforcer does not watch cannot be checked for integrity. Currently, the Enforcer takes no action if a malicious program accesses kernel memory or other critical resources In the absence of TCPA/TCG support for guaranteeing memory integrity, we are again forced to depend on the underlying OS to prevent such attacks.

Source Availability. The source code is available via GPL at enforcer. sourceforge.net. Our initial release went out in August 2003, with continual updates since. We brag that our project was the first non-trivial TCPA/TCG platform in the open world—but we brag cautiously, since (due to TCPA/TCG's intimate ties with product development, which tends to be held closely) one never knows. A the time of this writing, our project has had over 1150 source downloads, and still appears to the only open-source TCPA/TCG-based platform in existence.

11.5   Application: Hardened Apache

The first application we considered for this platform was revisiting the WebALPS application of Section 9.2.

The initial trust problem here is that, on the Web, server-side SSL only protects the tunnel between the client and the server; for sophisticated users, it might also provide some authentication of the server identity. How can we extend the secure tunnel into the server and around the application it offers? The solution approach was to move the SSL private key, management of the session keys, and the application into a TCP-housed co-server.

However, the limitations of the 4758 TCP hindered our 4758-based prototype. The Apache/mod_ssl/OpenSSL suite is popular for Web servers in part because it is free and it is relatively easy to set up and maintain. However, the 4758-based prototype required an expensive 4758 TCP. The implementation required sufficient careful surgery of the Apache codebase that updating our prototype for a new codebase was prohibitively hard. Moving a non-trivial application inside the platform required even more surgery. The performance was slow. Finally, had we managed to deploy this for a real Web site, we would have immediately been confronted with the lifetime issue: how often does the sysadmin go back to the SSL CA, and how can we help the SSL CA make less risky trust judgments about the future good behavior of the sysadmin?

Now that we had a TCP free from these constraints, it seemed natural to try the application again. We put the Apache suite in the medium-lived core, and the Web content and the SSL private key in the encrypted loopback filesystem; a symbolic link connected the pathname Apache likes to use for the key to its location within the loopback filesystem.

Performance. To gauge the impact of this TCP-hardened server, we gathered the static Web pages from the Dartmouth Athletics server (19K files, 664MB) and a typical day's access logs (20K URLs, 15% for files that did not exist),. The machine driving the tests was a dual-Intel Xeon machine, at 2Ghz with 512MB memory, running Debian Linux 2.4.20; the hardened Apache ran an IBM Netvista 8310 (Pentium 4, 2.00GHz, 128MB, Debian Linux kernel 2.6.0-test7, no preempt), The two machines were on 100Mb Ethernet, connected to the same switch.

When we built the configuration file for site, we modified 156 of the file hashes, just to check that Enforcer was working,

Our system—an Apache SSL server with all of the content in an encrypted loopback filesystem, using the TPM to protect the server's private key, and using the Enforcer for integrity— only had a 6.8% slowdown compared to a standard Apache SSL installation.

11.6   Application: OpenCA

Another part of our lab has focused on practical work in deploying a PKI in academic populations—and making it easier for other colleges and universities to do the same. In order to mesh with the current PKI-enabled applications (such as client-side SSL for Web authentication and authorization, EAP-TLS for WLAN access, and S/MIME for signing and encryption on email), such deployments must take the standard X.509 identity approach: each member of the enterprise generates a key pair, a central enterprise CA signs a certificate binding the public key to that individual's name, and all the applications implicitly trust that the enterprise CA knew what it was doing.

Deploying PKI in the higher education community raises several challenges, including:

- In order for stakeholders to have some assurance that an enterprise's CA certificates are meaningful, the CA's private key must be used only to sign bona fide certificates. This requirement implies the CA platform had better be trustworthy.

- In order to interact with a University B in the traditional X.509 way, the CA at University A must itself be certified by a CA that University B trusts. Typically, this second CA will either be a higher-level CA, or the CA of University B . If the latter, typically the two CAs *cross-certify* each other; a *bridge CA* exists exclusively to enable interaction by cross-certifying with other CAs.

  However, if a higher-level CA or a bridge CA is going to certify University A's CA, then this higher-level one needs assurance that University A follows appropriate certificate policies and practices, and also maintains a trustworthy platform for the CA. (Incidentally, our lab is currently setting up the *Higher Education Bridge CA (HEBCA)*, so we may have a chance to try these ideas in practice.)

- University IT departments typically underbudgeted, and so require a CA platform that is inexpensive.

- Like many enterprises, university IT departments typically are understaffed, and so require a CA platform that can remain trustworthy even without a dedicated specialist staff.

Using a TCPA/TCG-based TCP with the OpenCA software suite might address these problems. We use Bear/Enforcer to make sure that only the properly configured OpenCA platform can access the CA private key; as much as possible, we embody the practices and policies in the application installation itself, and use the TCP's outbound authentication feature to help increase the confidence of the certifying CA. Being based on a commodity machine and free software keeps the direct monetary cost low. Employing the physical protections of the TPM chip provides enhanced security. Finally, packaging this software—the TCP code coupled with OpenCA—in an easy-to-install "CA-in-a-box" would help keep manpower costs down.

As part of our lab's broader effort to produce "PKI-in-a-box" tools, we have been looking at a basic painless OpenCA tool. As part of his Master's thesis in hardening OpenCA, graduate student Josh Stabiner has integrated OpenCA with the Bear/Enforcer TCP code; an open-source release is in preparation.

11.7   Application: Compartmented Attestation

One of the most controversial aspects of turning a commodity machine into a TCP is its potential ability to give powerful corporations undue control over and access to consumer machines. In this vision, producers of a popular electronic resource, such as music files, may wish to limit what they perceive as illicit use of their intellectual property by ensuring that only consumer machines configured in an appropriately safe way can use these files.

For example, producer A might trust that program P, when installed in a properly configured system, might make it sufficiently difficult for an adversarial customer to make and distribute illicit copies of the licensed MP3s that A sells. Consequently, A might want a method to ensure that, before customer B buys an MP3 from A, B's machine is safely configured with P, and that A securely ships the MP3 to this program.

This scenario creates a conflict of interest. Given the permeable nature of modern operating systems, it would appear that A would need to know *everything* on B's machine, in order to be able to reasonably trust that the music was going to program P in a way that an adversarial B could not subvert. However, this knowledge violates B's privacy: even if B would like to engage in this license-restricted transaction with A, she might not want to share with A anything more than the fact she has P installed safely. She might very well regard the other items on her machine as none of A's business.

One way to overcome this problem is to build a TCP on B's machine with an operating system that can establish isolated *compartments* with reasonable assurance. B can set up P to run inside a compartment. The TCP's outbound authentication can convince A that it really is talking to an installation of P that, within the threat model, cannot be subverted by a malicious B; however, B can have assurance that A is learning the contents of the P compartment, but nothing else.

To implement this idea, we started with NSA's *Security-Enhanced Linux (SELinux)*. In theory, one can configure SELinux installations to enforce fairly strict security policies; in practice, the configuration process has considerable complexity. Nevertheless, we set it up to run the XMMS music player in a private compartment, and verified that even the root super-user cannot spy on its memory. We then modified our Bear/Enforcer code to work with the SELinux kernel, and to verify the security policy as well as the code suite. This task required getting the two LSMs—SELinux and Enforcer—to mesh together.

Besides providing an example of privacy-enhanced DRM, integrating SELinux with Bear/Enforcer also increases the trustworthiness of the platform, as SELinux policy enforcement will block many malicious attacks that could be attempted by code or users that still manage to get by the Enforcer's integrity check.

Student Alex Barsamian started building a prototype of this idea as part of his senior project. Graduate students Josh Stabiner and John Marchesini finished the prototype; an open-source release is in preparation.

As Section 12.2.5 discusses, recent work in secure hypervisors and trusted virtual machine monitors present yet another approach to compartmented attestation: expanding the "compartment" to be an entire virtual machine. The Terra team built a prototype based on insecure hardware [GPC+ 03]; another recent paper presents a paper design [SS03].

11.8   Further Reading

The basic Bear/Enforcer architecture was designed and developed with students John Marchesini and Omen Wild, based on the TCPA experiments carried out by student Rich MacDonald.

Our early technical reports [MSMW03, MSWM03] present the evolution of the project; Section 11.4 above was based on these reports. Our 2004 *ACSAC* paper [MSW+ 04] gives a presentation that focuses on the applications; Figure 11.1 above is a revised version of Figure 2 from our *ACSAC* paper.

As noted, our code exists at enforcer.sourceforge.com.

Chapter 12

NEW HORIZONS

Put simply, trusted computing platforms attempt to use some degree of hardware security to secure a broader platform and the distributed applications that use it. We have seen two main thrusts:

- secure coprocessors (trustworthy platforms protected by a physical security boundary), and

- trusted platform modules (smaller non-computational units that add trustworthiness to a platform that lies outside the physical security boundary).

(In this taxonomy, the areas of personal tokens and cryptographic accelerators would probably fit as an offshoot of the former.)

This chapter considers several current efforts that explore alternate technologies and approaches. We'll start by reviewing basic CPU privilege architecture and by framing the family of new paradigms that is emerging (Section 12.1). We'll consider academic research work in hardware techniques (Section 12.2) and in software and cryptographic techniques (Section 12.3). We'll then consider industrial efforts: both as currently available products (Section 12.4) and emerging product architectures (Section 12.5). We'll conclude in Section 12.6 by looking at the longer history of secure coprocessing in light of these new architectures.

12.1 Privilege Architectures

The textbook example of a processor has two privilege levels:

- a maximum privilege level, often called *supervisor* or *kernel* mode, and

- a lesser privilege level, often called *user* mode.

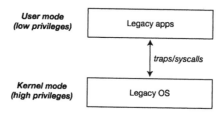

Figure 12.1. The standard CPU model offers two modes of execution: a low-privilege mode for user-level application code, and a higher-privilege mode for more sensitive operations, such as traditional OS functionality. The hardware constrains the interaction between modes, in order to help keep low-privilege code from accidentally or deliberately usurping higher privileges.

In this textbook example, the processor tracks the current privilege level via explicit hardware state. The intention is that the hardware itself will only carry out sensitive operations—such as working with the page tables controlling process memory—when the CPU is in kernel mode. Effectively exploiting these two privilege levels requires a way to move from low-privilege to high-privilege without negating the whole reason for two levels (as would happen if low-privilege code could simply ask to change to kernel mode). The processor typically provides a *trap* mechanism for controlled entry points into kernel mode: a trap instruction (and sometimes other hardware events) cause the CPU to change to kernel mode—but only after suspending the low-level thread of execution and moving instead to a separate and presumably more trustworthy thread.

In the textbook example, system software structure typically exploits this hardware base by mapping the OS to the supervisor level and the user-level processes to user level (hence the names for these levels). A wonderful example of hardware-based security, hardware restriction of sensitive operations can provide greatly increased protection of critical system resources against buggy and perhaps even malicious application code.

Figure 12.1 illustrates this architecture.

However, the above discussions focused on textbook examples. In reality, things are somewhat more complicated. For instance, on the hardware level, much of the computing world has standardized on the Intel x86 model, which has four levels of privilege—Ring 0 (maximum) to Ring 3 (minimum). However, one would be hard-pressed to find an example of a system that uses any levels other than Ring 0 and Ring 3. On the software level, some operating systems have operations that run at user level, and some permit user code at kernel level. Furthermore, operating systems tend to be large pieces of code that, historically, have numerous unintended ways for a malicious user to do far more than he or she is supposed to.

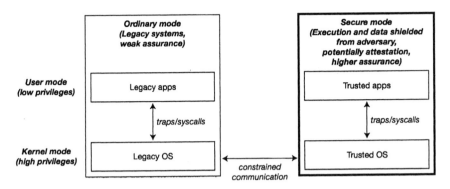

Figure 12.2. Emerging processor privilege architectures add another axis: a separate "secure" mode that provides increased protections against the adversary. Sensitive applications can be housed inside this protected half; other helper code executing inside this protected half may enhance overall system and application security through careful participation with the normal mode execution. Designs use various techniques to increase security by restricting interaction between the halves.

The New Paradigm. We will discuss many ongoing projects below, both in industry and in academia, are exploring various ways of enhancing CPU structure in order to make it easier to build trusted computing platforms. Viewing at a high level, we might characterize these enhancements as all variations of the same basic privilege architecture: revising the basic structure of Figure 12.1 by adding another axis. Besides "kernel" or "user," the CPU can also execute in "ordinary" mode or in "secure" mode. The exact terms differ among the projects, as do the implementation approaches. In "ordinary" mode, the processor acts like an ordinary, legacy CPU; in "secure" mode, it executes with higher assurances and stronger protections. To help ensure these protections hold, the architectures include techniques to manage and limit the execution entry points into the secure mode, and data movement into and out of this barrier.

Figure 12.2 illustrates this revised architecture.

12.2 Hardware Research

A secure coprocessor envelopes a multi-chip computing environment; a TPM protects a separate computing environment. Many academic researchers have explored alternative ways to add assurance to the CPU itself.

12.2.1 XOM

David Lie and his colleagues at Stanford proposed a CPU architecture based on *execute-only memory (XOM)* [LTM+ 00]. A driving motivation behind XOM was protecting against piracy and other adversarial manipulation of application software. In the resulting model, the application program may trust the CPU but not necessarily the OS. As in the encrypted storage model of Best and Kent and later prototyping by Yee (Chapter 4), the XOM architecture encrypts programs. Only the trusted processor possesses the necessary decryption key. Unlike the early prototypes, however, in XOM, only the sensitive application receives such protection. Hence, these encrypted compartments map to the "secure" mode in the new privilege paradigm.

The XOM design examines the CPU modifications necessary to make this idea work. The CPU possesses a private key; program ciphertext uses hybrid encryption, so that the CPU may employ faster symmetric cryptography for the bulk of its work. Cache lines and memory management include tagging, to bind storage data to XOM compartment, and keyed MACs for replay protection and integrity protection. These keyed MACs even depend on addresses, to protect against relocation attacks. The instruction set includes new hooks to enter and leave a protected XOM compartment, as well as to move data between a protected XOM compartment and an unprotected environment, and between the CPU and a XOM compartment's compartmentalized storage.

Subsequent XOM work included carrying out model checking to determine whether an adversarial OS could cause the system to enter insecure states [LMTH03]. This work demonstrated a replay attack that the literature had suggested, where an adversary could invalidate new data in the cache before it had been flushed to memory. Subsequent work also included exploring how to build an OS on top of this architecture [LTH03]—which required addressing questions such as how an OS can manage compartmentalized user processes without compromising them. As Lie and his colleagues worked through the various details of process and memory management, they made several modifications to the XOM hardware architecture. They also simulated the XOM CPU using SimOS [RBDH97] and built *XOMOS* by modifying IRIX 6.5 [SGI]. They benchmarked performance by running an MP3 player and OpenSSL.

To enable outbound authentication, Lie proposes using the program itself to carry its own private key in its protected code.

12.2.2   MIT AEGIS

Srini Devadas' group at MIT has also been active in enhancing the trustworthiness of CPUs. This work has resulted both in a new CPU architecture, *AEGIS* [SCG+03b], as well as numerous results in related supporting technologies. (The coincidence of this "AEGIS" name with the name of Arbaugh's secure boot project—Chapter 4—is unfortunate.)

Like the 4758, AEGIS starts with the assumption that the CPU possesses a secret the adversary cannot reach. Like XOM, AEGIS adds a new "secure" mode of execution to the CPU. AEGIS then adds new instruction set hooks to enter and exit this mode and to use this protected secret key. AEGIS looks at the security of memory for protected execution at a much finer granularity than

XOM; the initial design used hardware-accelerated Merkle hash trees to protect each access to each value, although additional techniques have been discussed. Like XOM, AEGIS provides all-or-nothing sharing: memory is either confined to a particular protected environment or not protected; unlike XOM, AEGIS implements this by partitioning the address space, instead of adding new data move instructions.

AEGIS includes two visions: when a trusted portion of the OS exists and when none exists; functionality moves from OS to hardware in the latter.

The AEGIS team speculates on applications in DRM and in *certified execution*: farming an application to a remote trusted computing platform that uses outbound authentication to prove it is "the real thing, doing the right thing." (TCP outbound authentication in this context is sometimes called *tamper-evident execution*.) The initial design was evaluated using simulation based on SimpleScalar tools, but silicon prototypes based on the OpenRISC core [Ope03, for example] are reportedly underway.

12.2.3 Cerium

Cerium is another trusted processor proposal from MIT [CM03]. Cerium consciously borrows architecture ideas from the 4758 and Dyad to support certified execution within a hardened CPU. Like XOM and AEGIS, Cerium does this by cryptographically protecting process address spaces. Cerium borrows the Merkle tree approach from the AEGIS team [SCG$^+$03a] (although the AEGIS team has also examined other schemes as well). Unlike XOM and AEGIS, Cerium uses software: it pins a trusted microkernel inside the CPU. Events that require address space manipulation trap to this microkernel, which directs the appropriate actions.

Cerium appears to have remained a paper design only.

12.2.4 Virtual Secure Coprocessing

Ruby Lee and her group at Princeton have also been looking at architectural changes to the CPU to enhance trustworthiness (as well as to further other security-related goals, such as faster cryptography). In contrast to the secure coprocessing and TPM approaches, Ruby suggested the term *virtual secure coprocessing* for approaches that build the necessary functionality into the CPU itself. One preliminary result included a design to create a safe space for cryptographic keys inside the device [ML02]. However, research efforts continue [LIB04, for example].

12.2.5 Virtual Machine Monitors

We could also try putting the security "underneath" an image of the machine, a well-studied idea in security [KZB$^+$91, for example] that is beginning to undergo a renaissance [The03, for example]. (Recall also Kent's foreshadowing of this idea in Section 4.1.1.)

In the research world, Tal Garfinkel and colleagues at Stanford have recently proposed *Terra* [GPC+ 03], an architecture which uses a *Trusted Virtual Machine Monitor* to produce separate virtual machines: both "open boxes" that act like ordinary open platforms, and "closed boxes" that provide some degree of trustworthiness. (Here, the "closed boxes" correspond to the secure mode in the new privilege paradigm.) The TVMM and underlying hardware would provide the security foundations and attestation hooks. Garfinkel and colleagues provide a thorough examination of the system design issues, and they even built a prototype on VMware GSX Server 2.0.1 with Debian GNU/Linux (although they note that this platform, not of "suitably high assurance," is for experimentation only). They have used this platform to build a "cheat-resistant" version of Quake. The Terra paper provides a lot of discussion relating that project both to previous trusted computing platform research (e.g., Chapter 4 and Chapter 7) as well as to emerging research (e.g., this chapter, Chapter 12).

In a similar vein, Leendert van Doorn at Watson has been leading a *secure hypervisor* research project, exploring using hardware support for machine virtualization, glued onto the TCPA/TCG architecture [Shy].

The reader should note that a subset of the security research community have very specific definitions for "virtualization"—and the VMware on x86 may not always qualify. [RI00] presents a good discussion of these issues.

Rumors exist that major industrial players may incorporate support for virtualization in future CPU hardware.

12.2.6   Others

XOM, AEGIS, Cerium, virtual secure coprocessing, and TVMMs all fall squarely in the space of research to directly transform a CPU into a TCP. However, much other exciting emerging research looks at other aspects of hardware-based security enhancements. We quickly review some.

Physical Unknown Functions. TCPs typically depend on hiding secrets from the adversary. As Chapter 3 discussed, the problem of hiding a secret in hardware—particularly low-cost chips—can be difficult. The AEGIS group at MIT has some intriguing results in using the random delays in circuit elements as a secret that (perhaps) can neither be reproduced nor "measured" except by direct execution of an untampered circuit [GCvD02].

Physical One-Way Functions. Another group at MIT has explored using epoxy with tiny glass spheres to build devices that calculate a function, determined by how these spheres deflect a laser shined at some angle, that (perhaps) can neither be reproduced nor simulated [PRTG02].

MEMSecurity. Bruce Donald and his group at Dartmouth have been exploring using *microelectromechanical systems (MEMS)* for security problems. MEMS are tiny devices (typically less than 100 microns) built of elements such as levers, gears, and springs; potential security applications include cryptography without the usual side-channel risks, as well as various hard-to-reproduce tokens.

HIDE. In a recent result, researchers at Georgia Tech have proposed *hardware support for leakage-Immune Dynamic Execution (HIDE)* [ZZPL03]. HIDE builds on the XOM model, and juggles code blocks between internal caches and external storage areas to try to keep the adversary from learning information about a protected execution from observing the address sequence. Further exploration of these techniques, in the spirit of oblivious RAM and practical private information retrieval, will be interesting.

SmashGuard. Carla Brodley and colleagues at Purdue have developed *Smash-Guard*, an approach that defends against buffer overflow attacks (Section 3.2.1) by modifying the CPU [OVB+04]. Such a hardware approach has the advantages of not degrading performance and not requiring re-compilation of legacy code.

CoPilot. Bill Arbaugh and his students at Maryland have developed *CoPilot*, which uses a hardware coprocessor and friendly DMA[1] to regularly verify the integrity of a desktop system [PFMA04]. Among the challenges here is working through exactly what constitutes integrity within the system and data components as they resides in memory.

TCG Extensions. Recently, Reiner Sailer and colleagues at IBM Watson have extended the TCPA/TCG architecture to observe and measure the dynamically changing state of a fully functional Linux platform [SZJv04]. This work required carefully wading through the components and dependencies that comprise such a complex, dynamic environment.

12.3 Software Research

Research efforts are also examining techniques not based on hardware but still relevant to trusted computing platforms.

---

[1] ...in contrast to the malicious DMA that might threaten a TCPA/TCG architecture

202   *TRUSTED COMPUTING PLATFORMS*

12.3.1   Software-based Attestation

A number of recent results have followed up on Bennet Yee's suggestion (Section 4.3) of using "behavior and timing checks" to determine integrity and authenticity of a remote machine.

Genuinity. In 2003, Rick Kennell and Leah Jamieson at Purdue considered the problem of how to determine whether a remote machine was indeed the "real thing doing the right thing," a concept they called *Genuinity* [KJ03]. In the Genuinity approach, the relying party challenges the machine to do something that involves many aspects of the machine's characteristics and state.

In 2004, Umesh Shankar and colleagues at Berkeley vigorously challenged these results, and demonstrated a series of attack techniques [SCT04].

SWATT. In 2004, Arvind Seshadri and colleagues at CMU presented *SWATT*, a similar software-based technique [SPLK04]. Unlike Genuinity, SWATT targets isolated embedded systems and avoids the attacks of Shankar.

12.3.2   Hiding in Plain Sight

Another branch of software techniques involve trying to run a sensitive program on the adversary's computer, but using techniques from cryptography and elsewhere to somehow provide TCP-like security, such as:

- assurance that the adversary cannot subvert this computation without detection, or

- assurance that the adversary cannot extract useful knowledge about the algorithm being carried out, or

- assurance that the adversary cannot extract useful knowledge about the parameters to this computation.

Research has provided a history of tantalizing results here, that (at a high level) might even seem contradictory. In 2001, Boaz Barak and his colleagues proved an impossibility result [BGI⁺01, Bar]. If we define an *obfuscater* as a program that takes another program as input and outputs an obfuscated version (that is functionally equivalent, but which an adversary cannot understand), then there exists a family of programs for which every obfuscater will fail to obfuscate. On the other hand, *encrypted functions* [ST98, for example] and *multi-party computation* [Gol04, Chapter 7, for example] show various promising results.

Paul van Oorschot also gives a good survey of protection techniques, including *white-box cryptography* and *software tamper resistance* in [van03].

12.4 Current Industrial Platforms

We quickly survey some interesting hardware techniques that are already part of the industrial base.

12.4.1 Crypto Coprocessors and Tokens

This book has treated the IBM 4758 as an exemplary trusted computing platform. However, as Chapter 5 observed, the 4758 became a commercial product not because IBM wanted a TCP, but because it wanted a flexible cryptographic accelerator.

Crypto accelerators and portable personal tokens often end up having several features that may make them appropriate for TCP applications. These features may include:

- a computational domain separate from the main processor;

- some degree of physical security;

- hardware support for cryptography;

- some amount of programmability.

As Moore's Law makes hardware smaller and cheaper, one might expect to see accelerators (such as devices from nCipher and Prism), as well as personal tokens (such as Dallas iButton, Fortezza card, Java cards, and USB tokens) look more and more like TCPs. Wave Systems has also been active in this space.

Indeed, IBM has recently produced a follow-on to the 4758, the *IBM 4764*, that preserves the same basic security architecture but with greatly increased computational and cryptographic power [Av04]. Unfortunately, as of this writing, developers toolkits are not available.

12.4.2 Execution Protection

As Section 3.2.1 discussed, code injection via overflowing a buffer on the stack of the victim machine remains a common source of compromise in the current information infrastructure. AMD [AMD04] and other major CPU vendors have started supporting execution protection in x86-class CPUs. A long-time feature of textbook examples, the *no execute (NX)* tag can prevent the CPU from executing code the adversary has injected. Both Linux and Windows are embracing this feature [Mim04, Roo04, for example].

However, many code-injection and attack techniques exist that do not require an executable stack. For example, consider the return-to-libc attack that Section 3.2.1 discussed, or the suite of techniques in [KLA$^+$04, pp. 191-196]. It will be interesting to see to see how NX-enhanced systems withstand adversarial scrutiny.

12.4.3   Capability-based Machines

Many old-timers wistfully recall *capability-based* and *tagged* architectures, exemplified by such systems as the CAP Computer at Cambridge [NW77] and the iAPX 432 architecture from Intel. Encapsulating an address with access rights, in an architecture that is aware of these rights at a fine granularity, has the potential for greatly increasing the trustworthiness of a computing platform.

Capability architectures are popularly regarded to have failed and vanished, even though the "failure" may have followed from other reasons, and even though current systems such as the AS/400 (and perhaps even the PowerPC) still support them.

At least one graybeard has suggested that perhaps it is time for a renaissance.

12.5   Looming Industry Platforms

As Chapter 10 discussed, the concepts of trusted computing have been percolating into standard industrial product design. A TPM that records configuration measurements and locks stored credentials to a particular suite of measurements is an add-on, peripheral to the main computing elements that occur on the platform. However, industrial efforts also loom that may bring trusted computing enhancements to these main elements as well.

On the hardware level, Intel and ARM have both been advancing enhancements to CPU architecture to bring increased protections into the processor itself. Section 12.5.1 considers Intel's *LaGrande* architecture and Section 12.5.2 considers ARM's *TrustZone*. On the software level, Microsoft's *Next Generation Secure Computing Base (NGSCB)* is an OS and application architecture that appears to build on and exploit such hardware foundations; Section 12.5.3 considers this effort.

As with Chapter 10, we need to include several caveats here. Since all of these projects are moving targets, this discussion will almost certainly be obsolete by the time it is read. Since this research is closely tied to potentially forthcoming products, the public availability of information (as well as actual hardware and software) is currently limited. Unlike TCPA/TCG, a university lab cannot yet simply download a specification and buy a machine and start experimenting.

12.5.1   LaGrande

Intel's LaGrande initiative is a security architecture intended to provide a hardware foundation for more trustworthy computing environments. This architecture encompasses many elements—including the CPU and the TCPA/TCG trusted platform module (version 1.2); according to rumors, forthcoming Intel chipsets may already possess these features, unactivated.

As is often heard about TCPA/TCG, LaGrande is often positioned as primarily defending a platform's user against software risks. The complexity of commodity operating systems, coupled with the myriad 3rd-party device drivers that must run at Ring 0 and the myriad channels for external adversaries to access an end-user system (and potentially exploit openings such as code-injection vulnerabilities), make the typical laptop or desktop a dangerous place to run sensitive code that uses sensitive data.

The driving force behind LaGrande appears to be to solve that problem, using hardware protections. Standard presentations focus on four main properties:

- **Protected Execution.** The platform needs a safe place for a sensitive application to run. Stakeholders need to be able to trust in this safety, even in the face of permeable, complex system software and potentially adversarial code.

- **Sealed Storage.** This application needs access to its own data. Stakeholders need to be able to trust in the secrecy (and presumably integrity) of this data, against the same threat model.

- **Attestation.** Remote relying parties need to be able to recognize that such a sensitive application running in such a trustworthy environment is indeed the "real thing, doing the right thing."

- **Protected I/O.** The machine in question primarily serves a local human user. This user and such applications need to a *trusted path* through which they can communicate. Keystroke sniffers are a well-known family of malware; crafting creative Web content that cause browsers to render misleading security information is a newer risk [FBDW97, YS02, YYS02].

It is interesting to note that, except "Protected I/O," these are the same properties we want for a trusted computing platform that can resist physical attack. As with TCPA/TCG, we might look at a LaGrande platform as another cousin of the 4758: far more power, but withstanding a different attack model. (It would also be interesting to examine what resistance against "logic analyzer" attacks might be possible, both with the current design and with hypothetical small enhancements, if one assumed the individual chips were secure against physical attack.)

The main elements of the LaGrande architecture are an enhanced CPU privilege model and the 1.2 TPM. LaGrande revises the basic structure of Figure 12.1 per the new privilege paradigm. Besides "kernel" or "user," the CPU can also execute as a *standard partition* or a *protected partition.* Apparently, there may many such partitions, and protected partitions provide more a trustworthy environment. The CPU will provide an instruction for securely switching protection contexts. A *domain manager* will handle these partitions; if on-chip code is required to implement the domain manager logic, the chip vendor must sign it, and the chip will verify the signature.

The TPM will assist with storing measurements and releasing credentials to the CPU only when the correct configuration is executing. (One can begin to see how the "localities" of the 1.2 TPM might work nicely with a new set of privilege modes in the CPU.)

This combination of enhanced CPU and TPM helps provide the protected execution, sealed storage, and attestation goals. Other avenues the LaGrande architecture appears to be considering are protection against malicious DMA, protection of the graphics card[2] (to prevent spoofing attacks—an idea that goes back at least to 1977 [Kar77, pages 95–100]), and ideas about other trusted paths. (From the NGSCB architecture—discussed below—one certainly suspects that LaGrande extends security mode tagging into the memory management unit.)

12.5.2 TrustZone

From a high level, *TrustZone* is a similar security architecture slowly emerging from ARM Holdings, which specializes in RISC architectures and drives a lot of the space for RISC microprocessors in embedded systems. Like the others, TrustZone adds the new axis to the basic privilege structure of Figure 12.1. Unlike LaGrande, TrustZone's approach is simpler: a single security bit indicates whether the CPU is in *normal* mode or *secure mode*. A *monitor* controls transitions; a new privileged instruction (*secure monitor interrupt*) traps to the controlled entry points in the monitor.

The monitor is intended to be small enough to make formal verification quite feasible. The monitor handles secure context switching; this task is made more efficient and simpler by having duplicate bookkeeping structures. A "security bit" also tags cache lines; normal and secure modes also have separate scratchpad memory. Other components in the system may also be aware of the normal/secure partition.

Thus, we start to see how TrustZone can provide protected execution and some sealed storage. The currently available literature does not yet document how stakeholder can be sure of which code is running inside the CPU in secure mode, nor how the code be sure of freshness, secrecy, and integrity of the data it stores outside the CPU.

12.5.3 NGSCB

A colleague of mine once observed he had never seen hardware do anything by itself; it usually requires software. Discussing details of new CPU architectures providing more trustworthy places for code begs the question of how software will take advantage of that. Similarly, discussing how a major hard-

---

[2]Researchers at Columbia University have recently begun examining what might be possible if the graphics processing unit is the *only* trusted module on a client machine [CBK04].

ware vendor is slowly incorporating this architecture into its chipsets begs the question of what market forces exist to cause that.

In our society's current computing environment, Microsoft dominates. It is natural to conclude that the flip side to the LaGrande architecture (and perhaps even to TCPA itself) has been the emerging Microsoft security architecture.

The architecture has been emerging slowly. Rumors blossomed into issued patents for a "Digital Rights Management Operating System" [EDL01a, EDL01b]. Popular press articles about a *Palladium* architecture [CJPL02], which seemed based on something like the then-TCPA architecture, only different, turned into the *Next Generation Secure Computing Base (NGSCB)*. Scholarly articles appeared [EP02, ELM$^+$03, for example], and more information emerged. The common belief was that NGSCB was going to be part of *Longhorn*, a new version of Windows slated for 2006; the Longhorn Developer Preview software development kit even included NGSCB code. Then, in Spring 2004, Microsoft appeared to be de-emphasizing NGSCB [Roo04, Fol04]. As of this writing, Microsoft's Web site lists the NGSCB documentation as "archived"; no current (non-archived) documentation is offered.

At a high level, one might understand NGSCB as an OS architecture that exploits the hardware features of LaGrande, while also positioning itself as a component of an OS with a huge body of legacy software and applications. Indeed, a feature/constraint that the literature touts repeatedly is the fact that NGSCB will not restrict what a user can run on her machine. Closing the environment to only authorized code (as we did with the IBM 4758) is not an option.

As with LaGrande, the NGSCB architecture advertises that it focuses on defending against the software attacks made possible by today's messy computing environments. ("This technology is not designed to provide defenses against hardware-based attacks that originate from someone in control of the local machine." [Mic03]) However, as we observed with LaGrande, it is tempting to consider how much further one could take this.

NGSCB stresses four key properties that map nicely to the four goals of LaGrande: "strong process isolation," "sealed storage," "trusted path to the user," and "attestation." NGSCB also uses a privilege model that maps to the quadrants of the new privilege paradigm (Figure 12.2). As is traditional, the user CPU mode (top row) holds application code and the kernel CPU mode (bottom row) holds kernel code. Of course, the messy state of current systems make this distinction weak, as we noted. NGSCB puts the messy current systems in the left column, and then (like LaGrande and TrustZone) adds a right column for a special secure mode. Some NGSCB documents even use *lhs* and *rhs* (for "left-hand side" and "right-hand side," respectively) to refer to these modes.

Unlike LaGrande and TrustZone, NGSCB spells out what software goes there:

- the *nexus* (a small security kernel, on the order of thousands of lines of code, not millions) goes in the secure-kernel quadrant; and

- *nexus computing agents (NCAs)* go in the secure-user quadrant.

NGSCB sketches suggest that multiple different *compartments* might swap into this right-hand side.

NGSCB expects the hardware will provide the appropriate functionality, including a *Security Support Component* (labeled as a 1.2 TPM in some diagrams) as well as possible CPU support for *curtained memory*: memory management that restricts RAM pages to the nexus-space only, and even denies DMA there.

12.6   Secure Coprocessing Revisited

It is interesting to consider the path we have traveled here. We started out by considering the construction and application of secure coprocessors: *separate* computational units co-located with a host that was less trustworthy. These coprocessors provide a trustworthy place for small, higher-assurance application entities that enhance the security of the host computation by cleverly participating in it. Figure 1.1 illustrated this architecture.

Limitations of this approach (as well as the commercial emergence of TCPA and TCG) led us to consider the role of a trusted platform module: a separate chip that lets us start thinking about the entire host machine as a (less) trusted computing platform. Academic projects—such as XOM, MIT AEGIS, and Terra—started moving protections based on a separate TPM to protections inside the CPU. Industrial projects also move protections inside the CPU. As Section 12.1 discussed, these emerging architectures add another axis to the standard privilege architecture. As Figure 12.2 sketched, the "normal" mode is open and accommodates legacy systems; the "secure" mode provides a more trustworthy place for small, higher-assurance application entities that enhance the security of the normal-mode computation by cleverly participating in it.

This looks just like secure coprocessing. Compare Figure 12.2 with Figure 1.1 back in Chapter 1. We map the normal mode to the host, the secure mode to the coprocessor, the nexus to the coprocessor OS, and the NCAs to the coprocessor applications.

How far can we take this analogy? How many of the coprocessor applications of Chapter 4, Chapter 9 and Chapter 11 will port to a LaGrande architecture? Can we take LaGrande ideas and move them back to coprocessors? What are the relative performance, flexibility, and security tradeoffs between a CPU approach, a TVMM approach, and a secure coprocessor approach?

In this book, we have charted the evolution of trusted computing platform ideas from decades ago to where they are now: on the cusp of merging with the dominant processor and operating system architectures. From one direction, we have seen an evolution of specialized coprocessors, that started with cryptographic accelerators and advanced to incorporate keys, operating systems, applications, and even PKIs. From the other direction, we have seen an evolution of techniques to add security to a general-purpose desktop. These evolutions are now converging.

What's going to happen next? In another decade, how many of these trusted computing platform design principles will be standard material in architecture and OS textbooks?

I cannot wait to see what the answers to these questions are.

12.7   Further Reading

[LTH03] probably gives the best overview of XOM, [SCG$^+$ 03b] of AEGIS, and [GPC$^+$ 03] of Terra.

[Lev84] gives a nice history of capability-based systems; although out of print, the book is available (for free) online.

For the emerging industrial architectures (Section 12.5), [Sta03] is an open-literature publication that provides a nice look at LaGrande. However, Intel is becoming increasingly open about the architecture, so more sources will undoubtedly emerge. [Hal03] provides a good overview of TrustZone. The archived material at Microsoft provide a good overview of NGSCB: the FAQ [Mic03] and the *Security Model* [Sec03] are particularly helpful.

GLOSSARY

AEGIS   Within the TCP space, two major projects have been named "AEGIS." Bill Arbaugh's doctoral work at Penn examined how to ensure that a standard machine boots securely (Section 4.3). More recently, Srini Devadas' group at MIT has examined how to incorporate protected spaces inside a CPU (Section 12.2.2).

AES   The *Advanced Encryption Standard* is a symmetric block cipher selected by the U.S. Government as the replacement to DES, as a strong, general purpose cipher that should not be vulnerable to brute-force attacks any time soon.

ANSI   The *American National Standards Institute* establishes standards for many things, including some cryptographic processes.

ARM   *Advanced RISC Machines* transformed into *ARM Holdings*, a firm that specializes in RISC-based CPU technology; ARM is the firm behind *TrustZone*. See Section 12.5.2.

asymmetric cryptography   *Public key cryptography* is also sometimes known as *asymmetric cryptography* due the asymmetry in its operations and keys.

BBRAM   In TCP design, *battery-backed RAM* is often a good place to store sensitive data: the batteries keep it non-volatile, but RAM is easier to zeroize than ROM or FLASH. In our IBM 4758 work, we ended up adopting the term "BBRAM."

Bellcore attack   The term *Bellcore attack* is another name for the *DFA* family of hardware attacks.

Beneš network   A Beneš Network is a compact circuit of $O(N \log N)$ binary switches that can generate any shuffle of its N inputs, depending on how the switches are set. See Section 9.4.4.

BIOS On modern computing systems, the first thing a machine executes at reset is *basic input/output system (BIOS)*: a fundamental, simple block of code.

bridge Much of PKI focuses on how Alice can construct a certificate chain from one of her trust roots to the signer of Bob's certificate. If an Alice and Bob are in separate enterprises, typically their enterprise CAs either cross-certify each other, or both become subordinate CAs to some higher-level one. In the cross-certification approach, N enterprises require $\Omega(N^2)$ cross-certifications. A way to simplify this arrangement is to have a designated *bridge* CA exist solely for cross-certification—so that this $\Omega(N^2)$ clique reduces to an $O(N)$ star.

CA In PKI, a *certification authority (CA)* is an entity who signs certificates binding public keys to something useful about the keyholder, such as their name. In popular use, the term "CA" usually implies a CA who issues X.509 identity certificates, in accordance with standard X.509 practices.

CCA The *Common Cryptographic Architecture (CCA)* is a family of cryptographic APIs used by many IBM customers. Consequently, CCA applications are one of the common uses of the the IBM 4758 platform. See Section 5.1.2.

Cerium *Cerium* is a project from MIT designed to support certified execution within a hardened CPU. See Section 12.2.3.

CP/Q++ CP/Q the embedded operating system we chose as the foundation for an OS layer for the IBM 4758 secure coprocessor. We deleted some modules and added others and, as a joke, called the result "CP/Q++." The joke became the official name.

cryptopaging Bennet Yee invented the concept of *cryptopaging*: using a host's memory as the backing store for the virtual memory system within a secure coprocessor. See Section 4.2.2.

DES The *data encryption standard (DES)* is a symmetric block cipher established by the U.S. government in 1975 and, after a time, widely adopted. Conventional wisdom still says the design is basically secure—except the 56-bit keyspace is now quite vulnerable to brute-force search. (DES was based the earlier Lucifer cipher from IBM.)

DFA *Differential Fault Analysis (DFA)* is a family of attacks on which the adversary induces an error in a device and then uses the subsequent incorrect operation to learn secrets. See Section 3.1.

digital signature In a *digital signature* scheme, Alice can calculate a *signature* from a message, such that only she can produce that signature, but anyone else can verify it.

DMA   *Direct Memory Access (DMA)* is a hardware technique that enables peripherals to access system memory without going through the CPU.

DPA   *Differential Power Analysis (DPA)* is an attack technique in which an adversary learns internal secrets of a device by using statistical methods on a set of power traces. See Section 3.3.2.

DRAM   One way to build RAM is to have an array of capacitors, one for each bit. The charge of the capacitor corresponds to the value stored at that bit. This *dynamic RAM (DRAM)* approach allows for inexpensive, space-efficient semiconductor memory—at the cost of having to continually read the bits and *refresh* them before the charge drains away.

DRM   Over history, commerce in physical media (such as books) implicitly came to depend on physical properties of these media—for example, it is easy to give a book away, but considerably harder to copy it. These physical restrictions led to rules of behavior that law and practice explicitly codified as *rights*—for example, consumer Alice may have the right to buy a book and then loan or sell it to Bob, but Alice may not have the right to make copies of this book and sell them. However, as media become digital, the physical restrictions no longer apply. *Digital Rights Management (DRM)* refers to the set of challenges—technical and otherwise—in trying to codify and enforce these usage rights in digital media.

DSA   The *Digital Signature Algorithm (DSA)* is a digital signature algorithm produced by the NSA and sanctioned by the U.S. government, as the officially blessed substitute for RSA. At the time, the motivations for a substitute appeared to step from both economics (DSA was intended to be free of patent encumbrances) as well as espionage (DSA only supported signatures; it did not support encryption).

Dyad   *Dyad* was a TCP project built by Bennet Yee and Doug Tygar at CMU using the IBM Citadel prototypes for secure coprocessing applications. See Section 4.1.4.

dynamic RAM   See *DRAM*.

EEPROM   *Electrically Erasable Programmable Read-Only Memory* is a type of non-volatile semiconductor memory that permits a limited number of updates, usually on a word basis.

endianness   Computer memory is typically organized as a linear array of bytes indexed by address. For a data item (such as a 32-bit integer) that requires a sequence of several bytes, the question arise: which *end* should come first? Architectures typically adopt a standard convention—e.g., "little endian" or "big endian."

FIPS   The *Federal Information Processing Standards (FIPS)* are a sequence of standards promulgated by NIST. Usually, a FIPS standard consists both of the set of rules in the standards document itself, as well as a process by which NIST can officially bless products as being compliant with those rules. FIPS address many cryptographic and security practices. In theory, U.S. law requires government players to purchase only FIPS-validated products.

FLASH   FLASH is a type of non-volatile semiconductor memory that can be erased and reprogrammed in the field. See Section 3.4.2.

FSM   The FIPS 140-N validation process required *finite state machine (FSM)* models of the platform. See Section 8.2.

Genuinity   In 2003, researchers from Purdue proposed a way to verify whether a remote system was indeed what we in the TCP community had been calling "the real thing doing the right thing." The Purdue approach presented the machine with a series of challenges and examined the values and timing of the responses. *Genuinity* was the name the researchers used for the general concept of "the real thing doing the right thing," but the community came to use the term to denote this specific approach. Follow-on work vigorously challenges the Purdue approach. See Section 12.3.1.

GPG   The *GNU Privacy Guard* is an open-source implementation of PGP.

hash   A *hash* function takes a message (usually arbitrary) to a fixed-length hash value. A *cryptographic hash function* usually is usually assumed to have other critical properties, such as being irreversible (it is computationally infeasible to find a message that maps to a given hash value) and *collision-resistant* (it is computationally infeasible to find two different messages that map to the same hash value). Although other types of hash functions exist in computer science, the universal presence of cryptographic hash functions in security work has eclipsed these other usages; say "hash" and everyone will assume "cryptographic hash." (Ralph Merkle has been quoted as saying that hash functions are the "duct tape of cryptography.")

HCISEC   *HCISEC* is an emerging research field that examines the interaction between usability and security.

HIDE   Researchers at Georgia Tech have recently proposed *hardware support for leakage-Immune Dynamic Execution (HIDE)*, a system to keep an adversary monitoring the address bus from learning what a CPU is doing. See Section 12.2.6.

HMAC   *Hash-based MAC (HMAC)* is a well-regarded way to build a keyed-MAC function from a hash function.

kernel mode   Conventional CPU privilege architecture offers two modes of execution; *kernel mode* is the name commonly given to the higher-privileged mode. See Section 12.1

LaGrande   *LaGrande* is the codename for a new Intel security architecture; the term is popularly used to refer both the architecture as well as to the chipset supporting that architecture. See Section 12.5.1.

Layer N In the IBM 4758 architecture, we divided software (and privileges) into a series of layers. Within this discussion of TCP, *Layer N* usually refers to a particular layer within that structure. See Section 5.4.

LBBRAM   In the IBM 4758 architecture, we used hardware locks to provide additional protections to some of the BBRAM. We termed this protected region *Lockable BBRAM (LBBRAM)*. See Section 6.4.3.

Level N The *FIPS 140-1* and *FIPS 140-2* standards for cryptographic modules provide an increasing series of security levels, from Level 1 (which some wags claim mean "the module possesses a manual") to Level 4. See Section 8.1.2.

LDAP   The *Lightweight Directory Access Protocol (LDAP)* is a standard protocol for providing (and accessing) directory services over the net.

LHS   Discussions of NGSCB often use *left-hand side (LHS)* to refer to the open, normal mode of CPU execution. See Figure 12.2 and Section 12.5.3.

LOCK   The *Logical Coprocessing Kernel (LOCK)* was a US Government project from the 1970s to use hardware to accelerate security operations in high-assurance computing systems. See Section 4.4.

Longhorn   *Longhorn* is the codename for a forthcoming version of Windows which, for a time, was reputed to include NGSCB.

LT   Discussions of LaGrande often use the acronym *LT* for "LaGrande technology."

MAC   In cryptography, a *message authentication code (MAC)* usually refers to a way of generating a code from a message that is infeasible without knowing a specific key. Because MACs are generally built from fast symmetric ciphers, I like to think of them as a "poor man's digital signature." The literature also sometimes calls these *keyed MACs*—apparently to remind the reader that a secret key is required. Some treatments also call them *Message Integrity Codes (MICs)*.

MEMS   *Microelectromechanical systems (MEMS)* are tiny physical systems built from things like gears and wires. Some researchers are currently examining MEMS as a new form of security hardware. See Section 12.2.6.

Miniboot   In the IBM 4758 platform, platform security was controlled by *Miniboot*, the software that ran before the OS booted. *Miniboot 0* resided in ROM; *Miniboot 1* resided in rewritable FLASH.

naive   In computer science parlance, a *naive* approach is one that everyone thought was natural and correct, until someone finally had a better insight. Consequently, the term does not usually have much of a pejorative connotation.

NGSCB   The *Next-Generation Secure Computing Base (NGSCB)* is the emerging Microsoft security architecture formerly called *Palladium*. See Section 12.5.3.

NIC   A *network interface card (NIC)* lies between a machine and its network. See Chapter 1.

non-volatile   *Non-volatile* storage does not lose its contents when device power is removed.

NX   New x86-class CPUs have introduced execution protection ("*NX*") that marks memory regions as not executable; both Windows and Linux have announced support. See Section 12.4.2.

oblivious circuits   One type of computation engine is a circuit, a set of of interconnected gates. The input values go in one side and the output comes out the other. In an *oblivious circuit*, the internal wires are encrypted and the gates each have some hidden control bits—so the adversary who cannot see inside the gates cannot know the details of the computation. We can sometimes use this technique to calculate secret functions on large data using small TCPs. See Section 9.4.4.

oblivious RAM   The term *oblivious RAM (ORAM)* refers a set of techniques, initially theoretical, by which a party can issue reads and writes to RAM while hiding information—particularly access patterns—from an adversary who monitors the buses.

opcode   An *opcode* is the actual binary sequence that embodies an executable instruction for a hardware processor.

ORAM   See *oblivious RAM*.

P2P   *Peer-to-peer (P2P)* applications and overlay networks have been a popular distributed computing paradigm in recent years. P2P systems try to make no distinction between client and server, and also try to incorporate principles of decentralization and self-organization.

Palladium   *Palladium* was the initial name for the Microsoft security initiative that subsequently was renamed *NGSCB*; by some accounts, a trademark issue triggered the name change.

PCR   In the TCPA/TCG architecture, a *platform configuration register (PCR)* stores measurements of system configuration. See Section 10.1.

permutation   A *permutation* is a bijective map from a set to itself. In our private information work, we use the term "permutation" to refer to this function on a set of integers that are indices of some other set; see *shuffle*.

PGP   Used mainly for signing and encrypting e-mail, *Pretty Good Privacy (PGP)* is a relatively unstructured approach to PKI where keyholders endorse other keyholder's keys, and can use fairly flexible "web of trust" policies to make trust judgments.

PIR *Private information retrieval (PIR)* is the initially theoretical problem of how a client can obtain a particular record from a server without the server knowing for which record the client asked. See Section 9.4.2.

PKI *Public key infrastructure (PKI)* refers to the technology necessary to use public key cryptography to solve trust problems in large populations. Many speakers use "PKI" to refer particularly to the problem of learning and verifying certificates for other parties (but in my lab, we like to take a broader view).

POST Many systems perform a *power-on self-test (POST)* at start-up; POST is usually so low-level that it runs before the operating system even boots and may even reside in ROM.

PPIR Dave Safford and I used the term *practical private information retrieval (PPIR)* to refer to solving the PIR problem (asking a server for a record, without the server knowing which one) while working within a real-world Web/SSL paradigm. See Section 9.4.

private key See *public key.*

PRNG A *pseudorandom number generator (PRNG)* generates numbers that are computationally indistinguishable from a truly random sequence. Generally, we use PRNGs in TCP design when we need to amplify a small number of random bits into a much larger sequence of bits that are random enough.

public key *Public key cryptography* cryptosystems generate a *key pair*, consisting of a *public key* and a *private key*, with the property that it is infeasible to calculate the private key from the public key. Because this approach to cryptography separates the privileges into two different keys, it enables many useful forms of trusted communication between parties that share no secrets a priori.

relying party In PKI parlance, a *relying party* is one who is trying to reach some trust decision about a particular public key.

RHS Discussions of NGSCB often use *right-hand side (RHS)* to refer to the closed, secure mode of CPU execution. See Figure 12.2 and Section 12.5.3.

Ring *N* The x86 CPU family provides four increasing levels of internal privilege, *Ring 0* (corresponding to kernel mode) through *Ring 3* (corresponding to user mode). *Ring 1* and *Ring 2* are seldom used. Some researchers use *Ring -1* to refer to a hypothetical privilege mode that is even more privileged than Ring 0.

RISC The *Reduced Instruction Set Computing (RISC)* approach to hardware gambles that by simplifying the instruction set, we can simplify the CPU and gain greater efficiency.

RNG A *random number generator (RNG)* generates random bits, unpredictable by the adversary.

RSA   *Rivest Shamir Adleman (RSA)* was the breakthrough public key cryptosystem, and remains the de facto universal standard.

RTM   In TCPA/TCG parlance, the *root of trust for measurement (RTM)* is the platform entity (e.g., BIOS) that the starts the platform configuration measurement process by measuring itself—and can thus subvert the whole process by lying. See Section 10.1.

SHA-1   The *secure hash algorithm, revision 1 (SHA-1)* is currently widely accepted as the cryptographic hash algorithm of choice.

Shibboleth   *Shibboleth* is a middleware system from Internet2 that enables an institution to share electronic resources over the Web with individuals from another institution, while abiding by inter-institutional access policy and without forcing a change in how individuals already authenticate to their home institutions. See Section 9.4.1.

shuffle   In our private information work, we use the term *shuffle* to refer to the rearrangement of an indexed set according to some permutation on its indices.

sHype   *sHype* is a secure hypervisor project housed at IBM Research. See Section 12.2.5.

SIDEARM   The *System-Independent Domain-Enforcing Assured Reference Monitor (SIDEARM)* was a hardware module inside the early *LOCK* system. See Section 4.4.

S/MIME   *S/MIME* is a format for incorporating encryption and digital signatures in electronic mail.

SPA   *Simple Power Analysis* is an attack technique in which an adversary learns internal secrets of a device by examining a power trace. See Section 3.3.2.

SRAM   See *static RAM*.

SSL   The *secure sockets layer (SSL)* protocol is a way for two parties to establish a secure tunnel over the network. Although intended to support general protocols, SSL is used in practice primarily to support secure connections from a Web browser to a Web server; in popular parlance, "SSL" refers exclusively to Web traffic served over an SSL channel.

static RAM   One way to build semiconductor RAM is to have an array of flip-flops, one for each bit. This approach requires more complex circuitry than dynamic RAM (since a flip-flop is significantly more than one capacitor), but does not require the complexity of refreshing.

SWATT   *SWATT* is a recent approach to verify whether a particular isolated embedded system is indeed "the real thing the right thing," by presenting challenges and observing the values and timings of responses. See Section 12.3.1.

symmetric key cryptography Before public key cryptography emerged in the 1970s, the only way to do cryptography required both parties to know the same secret key. *Symmetric key cryptography* (or sometimes just *symmetric crypto*) refers to this approach. Although it has more limited functionality than public key cryptography, symmetric cryptography still finds much use due to its higher performance.

TCG The *Trusted Computing Group (TCG)* is the consortium currently driving the TCPA/TCG architecture.

TCPA The *Trusted Computing Platform Alliance (TCPA)* is the now-defunct consortium that original drove the TCPA/TCG architecture. In popular parlance, the term *TCPA* is sometimes still used to refer to the architecture itself.

TCP A *trusted computing platform (TCP)* is a device that uses some degree of hardware enhancement to provide increased trustworthiness. See Section 1.1.

TDES Although now considered weak because its 56-bit keyspace is vulnerable to brute-force search, the *DES* symmetric block cipher is still considered otherwise more or less sound. Consequently, many applications iterate three rounds of DES to build a symmetric block cipher with a stronger keyspace (112 bits or 168 bits, depending on the construction). The term *triple DES (TDES)* refers to ciphers built this way. (However, now that *AES* has emerged, one expects to see uses of TDES giving way to AES.)

TEMPEST According to rumors, *TEMPEST* was a classified project that developed an extensive suite of defense technology against side-channel analysis. See Section 3.3.

Terra *Terra* is a recent design (and limited prototype) for building a TCP by securely virtualizing the CPU and machine. See Section 12.2.5.

TOCTOU A *time-of-check/time-of-use (TOCTOU)* vulnerability occurs when, during the duration between when a system tests a condition and when it acts on the results of that test, the condition may cease to hold. See Section 3.2.4.

TPM The heart of the TCPA/TCG architecture is a *trusted platform module (TPM)*, a separate chip from the CPU that provides a credential store keyed to platform configuration information. See Section 10.1.

TRM Steve Kent's doctoral research explored the use of what he termed *tamper-resistant modules (TRMs)*. See Section 4.1.1.

TSS The TCPA/TCG architecture includes a *TCG Software Stack (TSS)* that uses the TPM. A specification has recently been publicly released.

TVMM A *trusted virtual machine monitor (TVMM)* aspires to add security protections between the virtual machines it supports. See Section 12.2.5.

user mode Conventional CPU privilege architecture offers two modes of execution; *user mode* is the name commonly given to the lower-privileged mode. See Section 12.1

VMM A *virtual machine monitor (VMM)* supports multiple images of "virtual" machines. See Section 12.2.5.

volatile *Volatile* storage is not guaranteed to retain its contents when device power is removed.

X.509 *X.509* is a family of standards for handling public key certificates.

x86 The *x86* family, based on Intel designs, is probably the dominant CPU architecture on laptops and desktops today.

XOM *Execute-Only Memory (XOM)* is an approach to hardening CPUs by supporting executables that can only be decrypted inside an internal protected space. See Section 12.2.1.

zeroize *"Zeroize"* is the term the TCP field uses to refer to the rapid destruction of memory contents in the face of attack.

References

[AARR]   D. Agrawal, B. Archambeault, J. R. Rao, and P. Rohatgi. The EM Side-Channel(s): Attacks and Assessment Methodologies. Technical report, IBM Research. http://www.research.ibm.com/intsec/emf-paper.ps.

[AB96]   R. Anderson and S. Bezuidenhoudt. On the Reliability of Electronic Payment Systems. *IEEE Transactions on Software Engineering*, 22(5):294–301, 1996.

[ABF$^+$03] C. Aumüller, P. Bier, W. Fischer, P. Hofreiter, and J.-P. Seifert. Fault Attacks on RSA with CRT: Concrete Results and Practical Countermeasures. In *Cryptographic Hardware and Embedded Systems—CHES 2002*, pages 260–275. Springer-Verlag LNCS 2523, 2003.

[ADDS91]   D.G Abraham, G.M. Dolan, G.P. Double, and J.V. Stevens. Transaction security system. *IBM Systems Journal*, 30(2):206–229, 1991.

[AF03]   D. Asonov and J. Freytag. Almost Optimal Private Information Retrieval. In *Privacy Enhancing Technologies—PET 2002*, pages 209–223. Springer-Verlag LNCS 2482, 2003.

[AFS97]   W. Arbaugh, D. Farber, and J. Smith. A Secure and Reliable Bootstrap Architecture. In *Proceedings of the 1997 Symposium on Security and Privacy*, pages 65–71. IEEE, 1997.

[AK96]   R. Anderson and M. Kuhn. Tamper Resistance—A Cautionary Note. In *Proceedings of the 2nd USENIX Workshop on Electronic Commerce*, pages 1–11, 1996.

[AK97]   R. Anderson and M. Kuhn. Low Cost Attacks on Tamper Resistant Devices. In *Proceedings of the 1997 Security Protocols Workshop*, pages 125–136. Springer-Verlag LNCS 1361, 1997.

[AMD04]   AMD and Microsoft to Provide Customers New Security Technology. Press release, February 2004.

[Anda]   R. Anderson. TCPA/Palladium Frequently Asked Questions. http://www.cl.cam.ac.uk/users/rja14/tcpa-faq.html.

[Andb]   R. Anderson. Two Remarks on Public Key Cryptology. `http://www.ftp.cl.cam.ac.uk/ftp/users/rja14/forwardsecure.pdf`. This write-up documents Ross's invited lecture at the *ACM Conference on Computer and Communications Security* in 1997, and is regarded as the seminal work in forward security in public-key cryptography.

[And01]   R. Anderson. *Security Engineering: A Guide to Building Dependable Distributed Systems*. John Wiley & Sons, 2001.

[ARR03]   D. Agrawal, J.R. Rao, and P. Rohatgi. Multi-channel Attacs. In *Cryptographic Hardware and Embedded Systems—CHES 2003*, pages 2–16. Springer-Verlag LNCS 2779, 2003.

[Asn04]   D. Asnonov. *Querying Databases Privately: A New Approach to Private Information Retrieval*. Springer-Verlag LNCS 3128, 2004.

[AUH99]   C. Antonelli, M. Undy, and P. Honeyman. The Packet Vault: Secure Storage of Network Data. In *Proceedings of the Workshop on Intrusion Detection and Network Monitoring*, pages 103–109. USENIX, 1999.

[Av04]   T. Arnold and L. van Doorn. The IBM PCIXCC: A New Cryptographic Co-processor for the IBM eServer. *IBM Journal of Research and Development*, 48:475–487, 2004.

[BA01]   M. Bond and R. Anderson. API-Level Attacks on Embedded Systems. *IEEE Computer*, 34:64–75, October 2001.

[Bar]   B. Barak. Can We Obfuscate Programs? `http://www.math.ias.edu/~boaz/Papers/obf_informal.html`. An informal, accessible discussion of [BGI$^+$01].

[BB03]   D. Brumley and D. Boneh. Remote Timing Attacks Are Practical. In *Proceedings of the 12th USENIX Security Symposium*, pages 1–14, 2003.

[BBC$^+$00]   S. Beattie, A. Black, C. Cowan, C. Pu, and L. Yang. CryptoMark: Locking the Stable door ahead of the Trojan Horse. White Paper, WireX Communications Inc., 2000.

[BDL97]   D. Boneh, R.A. DeMilllo, and R.J. Lipton. On the importance of checking cryptographic protocols for faults. In *Advances in Cryptology, Proceedings of EUROCRYPT '97*, pages 37–51. Springer-Verlag LNCS 1233, 1997. A revised version appeared in the *Journal of Cryptology* in 2001.

[BDTW01]   D. Boneh, X. Ding, G. Tsudik, and C.M. Wong. A Method for Fast Revocation of Public Key Certificates and Security Capabilities. In *Proceedings of the 10th USENIX Security Symposium*, pages 297–308, 2001.

[BECN$^+$04] H. Bar-El, H. Choukri, D. Naccache, M. Tunstall, and C. Whelan. The Sorcerer's Apprentice Guide to Fault Attacks. In *Workshop on Fault Detection and Tolerance in Cryptography*, 2004. `http://www.gemplus.com/smart/r_d/publications/pdf/BCN_04sor.pdf`.

[Bes80]   R. Best. Preventing Software Piracy with Crypto-Microprocessors. In *Proceedings of the IEEE Spring Compcon 80*, pages 466–469, 1980.

[BGI+01]   B. Barak, O. Goldreich, R. Impagliazzo, S. Rudich, A. Sahai, S. Vadhan, and K. Yang. On the (Im)possibility of Obfuscating Programs. In *Advances in Cryptology—Crypto 01*, pages 1–18. Springer-Verlag LNCS 2139, 2001.

[CBK04]   D. Cook, R. Baratto, and A. Keromytis. Remotely Keyed Cryptographics: Secure Remote Display Access Using (Mostly) Untrusted Hardware. http://www1.cs.columbia.edu/~dcook/pubs/rkey050504.pdf, May 2004. Manuscript.

[Cha84]   D. Chaum. Design Concepts for Tamper Responding Systems. In *Advances in Cryptology—Proceedings of Crypto 83*, pages 387–392. Plenum, 1984.

[Cha85]   D. Chaum. Security without Identification: Transaction Systems to Make Big Brother Obsolete. *Communications of the ACM*, 28(10):1030–1044, 1985.

[CJPL02]   A. Carroll, M. Juarez, J. Polk, and T. Leininger. Microsoft "Palladium": A Business Overview. Microsoft PressPass, August 2002.

[CLRS01]   T. "BBQ" Cormen, C. Leiserson, R. Rivest, and C. Stein. *Introduction to Algorithms*. McGraw-Hill, 2nd edition, 2001.

[Clu03]   J.S. Clulow. The Design and Analysis of Cryptographic APIs for Security Devices. Master's thesis, University of Natal, Durban, South Africa, 2003.

[CM03]   B. Chen and R. Morris. Certifying Program Execution with Secure Processors. In *9th Hot Topics in Operating Systems (HOTOS-IX)*, 2003.

[Com04]   Common Criteria for Information Technology Security Evaluation. Version 2.2, Revision 256, CCIMB-2004-01-001, January 2004.

[CW96]   E. Clarke and J. Wing. Formal Methods: State of the Art and Future Directions. *ACM Computing Surveys*, 28:626–643, 1996.

[Dep85]   Department of Defense Trusted Computer System Evaluation Criteria. DoD 5200.28-STD, December 1985. This is better known as the "Orange Book," due to the color of the cover on the hardcopy.

[DLP+01]   J. Dyer, M. Lindemann, R. Perez, R. Sailer, S.W. Smith, L.van Doorn, and S. Weingart. Building the IBM 4758 Secure Coprocessor. *IEEE Computer*, 34:57–66, October 2001.

[DPSL99]   J. Dyer, R. Perez, S.W. Smith, and M. Lindemann. Application Support Architecture for a High-Performance, Programmable Secure Coprocessor. In *22nd National Information Systems Security Conference*, October 1999.

[EDL01a]   P. England, J. DeTreville, and B. Lampson. Digital Rights Management Operating System, December 2001. United States Patent 6,330,670.

[EDL01b]   P. England, J. DeTreville, and B. Lampson. Loading and Identifying a Digital Rights Management Operating System, December 2001. United States Patent 6,327,652.

[ELM+03]   P. England, B. Lampson, J. Manferdelli, M. Peinado, and B. Willman. A Trusted Open Platform. *IEEE Computer*, pages 55–62, July 2003.

[EP02]   P. England and M. Peinado. Authenticated Operation of Open Computing De-
            vices. In *Information Security and Privacy*, pages 346–361. Springer-Verlag
            LNCS 2384, 2002.

[FBDW97]   E. Felten, D. Balfanz, D. Dean, and D. Wallach. Web Spoofing: An Internet
            Con Game. In *20th National Information Systems Security Conference*, 1997.

[Fel03]   E. Felten. Understanding Trusted Computing. *IEEE Security and Privacy*,
            pages 60–62, May/June 2003.

[Fol04]   M. Foley. Microsoft: 'Palladium' Is Still Alive and Kicking. Microsoft Watch,
            May 2004.

[GA03]   S. Govindavajhala and A.W. Appel. Using Memory Errors to Attack a Virtual
            Machine. In *Proceedings of the 2003 Symposium on Security and Privacy*,
            pages 154–165. IEEE, 2003.

[GCvD02]   B. Gassend, D. Clarke, M. van Dijk, and S. Devadas. Silicon Physical Ran-
            dom Functions. In *Proceedings of the 9th ACM Conference on Computer and
            Communications Security*, pages 148–160, 2002.

[GGKL89]   M. Gasser, A. Goldstein, C. Kaufman, and B. Lampson. The Digital Distributed
            System Security Architecture. In *Proceedings of the 12th NIST-NCSC National
            Computer Security Conference*, pages 305–319, 1989.

[GO96]   O. Goldreich and R. Ostrovsky. Software Protection and Simulation on Obliv-
            ious RAMs. *Journal of the ACM*, 43(3):431–473, 1996.

[Gol04]   O. Goldreich. *Foundations of Cryptography: Volume 2, Basic Applications*.
            Cambridge University Press, 2004.

[GPC+03]   T. Garfinkel, B. Pfaff, J. Chow, M. Rosenblum, and D. Boneh. Terra: A Virtual
            Machine-Based Platform for Trusted Computing. In *Proceedings of the 19th
            ACM Symposium on Operating Systems Principles (SOSP 2003)*, pages 193–
            206, 2003.

[GSTY96]   H. Gobioff, S.W. Smith, J.D. Tygar, and B.S. Yee. Smart Cards in Hostile
            Environments. In *Proceedings of the 2nd USENIX Workshop on Electronic
            Commerce*, pages 23–28, 1996.

[Gun90]   C. Gunther. An Identity-Based Key-Exchange Protocol. In *Advances in
            Cryptology—Eurocrypt '89*, pages 29–37. Springer-Verlag LNCS 434, 1990.

[Gut96]   P. Gutmann. Secure Deletion of Data from Magnetic and Solid-State Memory.
            In *Proceedings of the 6th USENIX Security Symposium*, pages 77–89, 1996.

[Gut01]   P. Gutmann. Data Remanence in Semiconductor Devices. In *Proceedings of
            the 10th USENIX Security Symposium*, pages 39–54, 2001.

[Gut04]   P. Gutmann. *Cryptographic Security Architecture: Design and Verification*.
            Springer-Verlag, 2004.

[Hal03]   T. Halfhill. ARM Dons Armor: TrustZone Security Extensions Strengthen
            ARMv6 Architecture. Microprocessor Report 8/25/03-01, August 2003.

[HG04]   A. Herzberg and A. Gbara. Protecting (even) Naive Web Users, or: Preventing Spoofing and Establishing Credentials of Web Sites. Cryptology ePrint Archive, Report 2004/155, 2004. http://eprint.iacr.org/.

[Hig86]   H.J. Highland. Electromagnetic Radiation Revisited. *Computers and Security*, 5:85–100, 1986.

[HKK93]   H. Härtig, O. Kowalski, and W. Kühnhauser. The BirliX Security Architecture. *Journal of Computer Security*, 2(1):5–21, 1993.

[HMMW95]   W. Havener, R. Medlock, R. Mitchell, and R. Walcott. *Derived Test Requirements for FIPS PUB 140-1*. National Institute of Standards and Technology, 1995.

[IAPR02]   N. Itoi, W. Arbaugh, S. Pollack, and D. M. Reeves. Personal Secure Booting. In *Information Security and Privacy*, pages 130–144. Springer-Verlag LNCS 2384, 2002.

[IBM]   IBM Watson Global Security Analysis Lab. TCPA Resources. http://www.research.ibm.com/gsal/tcpa.

[IS03a]   A. Iliev and S.W. Smith. Privacy-Enhanced Credential Services. In *2nd Annual PKI Research Workshop*. NIST/NIH/Internet2, April 2003.

[IS03b]   A. Iliev and S.W. Smith. Prototyping an Armored Data Vault: Rights Management for Big Brother's Computer. In *Privacy Enhancing Technologies—PET 2002*, pages 144–159. Springer-Verlag LNCS 2482, 2003.

[IS04a]   A. Iliev and S.W. Smith. Enhancing User Privacy via Trusted Computing at the Server: Two Case Studies. *IEEE Security and Privacy*, 2004. Accepted for publication.

[IS04b]   A. Iliev and S.W. Smith. Private Information Storage with Logarithmic-space Secure Hardware. In *Information Security Management, Education, and Privacy*, pages 201–216. Kluwer, 2004.

[Ito00]   N. Itoi. Secure Coprocessor Integration with Kerberos V5. In *Proceedings of the 9th USENIX Security Symposium*, pages 113–128, 2000.

[Jia01]   S. Jiang. WebALPS Implementation and Performance Analysis: Using Trusted Co-servers to Enhance Privacy and Security of Web Interactions. Master's thesis, Dartmouth College Department of Computer Science, June 2001.

[JSM01]   S. Jiang, S.W. Smith, and K. Minami. Securing Web Servers against Insider Attack. In *Seventeenth Annual Computer Security Applications Conference*, pages 265–276. IEEE Computer Society, 2001.

[KA98]   M. Kuhn and R. Anderson. Soft Tempest: Hidden Data Transmission Using Electromagnetic Emanations. In *Information Hiding 1998*, pages 124–142. Springer-Verlag LNCS 1525, 1998.

[Kar77]   P. Karger. Non-Discretionary Access Control for Decentralized Computing Systems. Master's thesis, Massachusetts Institute of Technology Laboratory for Computer Science, 1977. Available as Technical Report MIT/LCS/TR-199; see http://ncstrl.mit.edu/.

[Ken80] S. Kent. *Protecting Externally Supplied Software in Small Computers.* PhD thesis, Massachusetts Institute of Technology Laboratory for Computer Science, 1980.

[KJ03] R. Kennell and L. Jamieson. Establishing the Genuinity of Remote Computer Systems. In *Proceedings of the 12th USENIX Security Symposium*, pages 295–308, 2003.

[KJJ99] P. Kocher, J. Jaffe, and B. Jun. Differential Power Analysis. In *Advances in Cryptology—Crypto 99*. Springer-Verlag LNCS 1666, 1999.

[KLA+04] J. Koziol, D. Litchfield, D. Aitel, C. Anley, S. Eren, N. Mehta, and R. Hassell. *The Shellcoder's Handbook: Discovering and Exploiting Security Holes.* Wiley, 2004.

[KM97] M. Kaufmann and J. S. Moore. An Industrial Strength Theorem Prover for a Logic Based on Common Lisp. *IEEE Transactions on Software Engineering*, 23(4):203–213, 1997.

[KM00] R. Kohlas and U. Maurer. Reasoning about Public-Key Certification: On Bindings Between Entities and Public Keys. *Journal on Selected Areas in Communications*, 18:551–560, 2000.

[Kno01] E. Knop. Secure Public-Key Services for Web-Based Mail, August 2001. Senior Thesis, Dartmouth College Department of Computer Science.

[Koc96] P. Kocher. Timing Attacks on Implementations of Diffie-Hellman, RSA, DSS, and Other Systems. In *Advances in Cryptology—Crypto 96*. Springer-Verlag LNCS 1109, 1996.

[KS74] P. Karger and R. Schell. MULTICS Security Evaluation: Vulnerability Analysis. Technical Report EDS-TR-74-193, Vol II, Hanscom AFB, Electronic Systems Division (AFSC), 1974.

[KS94] G. Kim and E. Spafford. The Design and Implementation of Tripwire: a File System Integrity Checker. In *Proceedings of the 2nd ACM Conference on Computer and Communications Security*, pages 18–29. ACM, ACM Press, 1994.

[Kuh02] M. Kuhn. Optical Time-Domain Eavesdropping Risks of CRT Displays. In *Proceedings of the 2002 Symposium on Security and Privacy*, pages 3–18. IEEE, 2002.

[Kuhar] M. Kuhn. Electromagnetic Eavesdropping Risks of Flat-Panel Displays. In *Privacy Enhancing Technologies, Fourth International Workshop*. Springer-Verlag LNCS, To appear.

[KZB+91] P. Karger, M. Zurko, D. Bonin, A. Mason, and C. Kahn. A Retrospective on the VAX VMM Security Kernel. *IEEE Transactions on Software Engineering*, 17:1147–1165, 1991.

[LABW92]   B. Lampson, M. Abadi, M. Burrows, and E. Wobber. Authentication in Distributed Systems: Theory and Practice. *ACM Transactions on Computer Systems*, 10(4):265–310, 1992.

[LBK04]   M. Lorch, J. Basney, and D. Kafura. A Hardware-secured Credential Repository for Grid PKIs. In *4th IEEE/ACM International Symposium on Cluster Computing and the Grid*, 2004.

[Lev84]   H. Levy. *Capability-Based Computer Systems*. Digital Press, 1984. Out of print, but a free online copy lives at http://www.cs.washington.edu/homes/levy/capabook/.

[LIB04]   R. Lee, C. Irvine, and T. Benzel. Research Agenda for Unified Core Mechanisms in Highly Secure Mobile Platforms. In *Security Challenges at the Foundation: Secure Computing Enabled by Hardware, Firmware and Low-Level Software*. DARPA Invitational Workshop, 2004.

[LMTH03]   D. Lie, J. Mitchell, C. Thekkath, and M. Horowitz. Specifying and Verifying Hardware for Tamper-Resistant Software. In *Proceedings of the 2003 Symposium on Security and Privacy*, pages 166–177. IEEE, 2003.

[LR88]   M. Luby and C. Rackoff. How to Construct Pseudo-Random Permutations from Pseudo-Random Functions. *SIAM Journal on Computing*, 17(2):373–386, 1988.

[LS01]   M. Lindemann and S.W. Smith. Improving DES Coprocessor Throughput for Short Operations. In *Proceedings of the 10th USENIX Security Symposium*, pages 67–81, August 2001.

[LTH03]   D. Lie, C. Thekkath, and M. Horowitz. Implementing an Untrusted Operating System on Trusted Hardware. In *Proceedings of the 19th ACM Symposium on Operating Systems Principles (SOSP 2003)*, pages 178–192, 2003.

[LTM+00]   D. Lie, C. Thekkath, M. Mitchell, P. Lincoln, D. Boneh, J. Mitchell, and M. Horowitz. Architectural Support for Copy and Tamper Resistant Software. In *Proceedings of the 9th International Conference on Architectural Support for Programming Languages and Operating Systems—ASPLOS-IX*, pages 168–177, 2000.

[Mar04]   J. Marchesini. HEMP: Hardware-Enhanced MyProxy, June 2004. Ph.D. thesis proposal, Dartmouth College Department of Computer Science.

[Mau96]   U. Maurer. Modelling a Public-Key Infrastructure. In *Computer Security— ESORICS 96*, pages 325–350. Springer-Verlag LNCS 1146, 1996.

[Mic03]   Microsoft Next-Generation Secure Computing Base—Technical FAQ. Microsoft TechNet, July 2003.

[Mim04]   M. Mimoso. NX Slams Door on Linux Buffer Exploits. SearchEnterpriseLinux.Com, June 2004.

[ML02]   P. McGregor and R. Lee. Virtual Secure Co-Processing on General-purpose Processors. Technical Report CE-L2002-003, Princeton University, November 2002.

[MS02]   J. Marchesini and S.W. Smith. Virtual Hierarchies: An Architecture for Building and Maintaining Efficient and Resilient Trust Chains. In *Proceedings of the 7th Nordic Workshop on Secure IT Systems—NORDSEC 2002*. Karlstad University Studies, November 2002.

[MSMW03]   R. Macdonald, S.W. Smith, J. Marchesini, and O. Wild. Bear: An Open-Source Virtual Secure Coprocessor based on TCPA. Technical Report TR2003-471, Dartmouth College Department of Computer Science, August 2003.

[MSW⁺04]   J. Marchesini, S.W. Smith, O. Wild, J. Stabiner, and A. Barsamian. Open-Source Applications of TCPA Hardware. In *20th Annual Computer Security Applications Conference*. IEEE Computer Society, December 2004. To appear.

[MSWM03]   J. Marchesini, S.W. Smith, O. Wild, and R. Macdonald. Experimenting with TCPA/TCG Hardware, Or: How I Learned to Stop Worrying and Love The Bear. Technical Report TR2003-476, Dartmouth College Department of Computer Science, December 2003.

[Nat94]   National Institute of Standards and Technology. *Security Requirements for Cryptographic Modules*. Federal Information Processing Standards Publication 140-1, 1994.

[Nat01]   National Institute of Standards and Technology. *Security Requirements for Cryptographic Modules*. Federal Information Processing Standards Publication 140-2, 2001.

[Neu95]   P. Neumann. *Computer-Related Risks*. Addison-Wesley, 1995.

[NTW01]   J. Novotny, S. Tueke, and V. Welch. An Online Credential Repository for the Grid: MyProxy. In *Proceedings of the 10th International Symposium on High Performance Distributed Computing (HPDC-10)*, pages 104–111. IEEE, 2001.

[NW77]   R.M. Needham and R. Walker. The Cambridge CAP Computer and its Protection System. In *Proceedings of the 6th Symposium on Operating System Principles*, pages 1–10, 1977.

[Ope03]   OpenRISC 1000 Architecture Manual. OPENCORES.ORG, 2003.

[OVB⁺04]   H. Ozdoganoglu, T. Vijaykumar, C. Brodley, A. Jalote, and B. Kuperman. SmashGuard: A Hardware Solution to Prevent Security Attacks on the Function Return Address. Technical Report TR-ECE 03-13, Purdue University Electrical and Computer Engineering, 2004.

[Pal92]   E. Palmer. An Introduction to Citadel—A Secure Crypto Coprocessor for Workstations. Technical Report RC18373, IBM T.J. Watson Research Center, 1992.

[Pea03]   S. Pearson, editor. *Trusted Computing Platforms: TCPA Technology in Context*. Prentice Hall, 2003.

[Per03]   M. Periera. Trusted S/MIME Gateways, May 2003. Senior Honors Thesis, Dartmouth College Department of Computer Science.

[PFMA04]   N. Petroni, T. Fraser, J. Molina, and W.A. Arbaugh. Copilot—a Coprocessor-based Kernel Runtime Integrity Monitor. In *Proceedings of the 13th USENIX Security Symposium*, pages 179–194, 2004.

[PH04]   R. Phan and H. Handschuh. On Related-Key and Collision Attacks: The Case
            for the IBM 4758 Cryptoprocessor. In *Information Security: 7th International
            Conference, ISC 2004*, pages 111–122. Springer-Verlag LNCS 3225, 2004.
            Unfortunately, the authors missed the fundamental point that the 4758 platform
            is *not* the same thing as the CCA application.

[Pri86]   W. L. Price. Physical Security of Transaction Devices. Technical Report DITC
            4/86, National Physical Laboratory, 1986.

[PRTG02]   R. Pappu, B. Recht, J. Taylor, and N. Gershenfeld. Physical One-Way Functions.
            Science, 297:2026–2030, 2002.

[PSST02]   A. Perrig, S.W. Smith, D. Song, and J.D. Tygar. SAM: A Flexible and Se-
            cure Auction Architecture using Trusted Hardware. *eJETA.org: The Electronic
            Journal for E-Commerce Tools and Applications*, 1, January 2002.

[QS01]   J.-J. Quisquater and D. Samyde. ElectroMagnetic Analysis (EMA): Measures
            and Countermeasures for Smart Cards. In *Smart Card Programming and Secu-
            rity*, pages 200–210. Springer-Verlang LNCS 2140, 2001.

[QS02]   J.-J. Quisquater and D. Samyde. Side channel cryptanalysis. In *Atelier SEcurité
            des Communications sur Internet (SECI'02)*, pages 179–184, 2002.

[RBDH97]   M. Rosenblum, E. Bugnion, S. Devine, and S. Herrod. Using the SimOS Ma-
            chine Simulator to Study Complex Computer Systems. *Modeling and Computer
            Simulation*, 7(1):78–103, 1997.

[Res00]   E. Rescorla. *SSL and TLS: Designing and Building Secure Systems*. Addison-
            Wesley, 2000.

[RG91]   D. Russell and G.T. Gangemi. *Computer Security Basics*. O' Reilly, 1991.

[RI00]   J. Robin and C. Irvine. Analysis of the Intel Pentium's Ability to Support a
            Secure Virtual Machine Monitor. In *Proceedings of the 9th USENIX Security
            Symposium*, 2000.

[Roo04]   P. Rooney. Microsoft Shelves NGSCB Project As NX Moves To Center Stage.
            CRN, May 2004.

[SA98]   S.W. Smith and V. Austel. Trusting Trusted Hardware: Towards a Formal Model
            for Programmable Secure Coprocessors. In *Proceedings of the 3rd USENIX
            Workshop on Electronic Commerce*, August 1998.

[SA03]   S. Skorobogatov and R. Anderson. Optical Fault Induction Attacks. In *Crypto-
            graphic Hardware and Embedded Systems—CHES 2002*, pages 2–12. Springer-
            Verlag LNCS 2523, 2003.

[Saf02a]   D. Safford. Clarifying Misinformation on TCPA. http://www.research.
            ibm.com/gsal/tcpa/tcpa_rebuttal.pdf, October 2002.

[Saf02b]   D. Safford. The Need for TCPA. http://www.research.ibm.com/gsal/
            tcpa/why_tcpa.pdf, October 2002.

[SAH00]   S.W. Smith, C. Antonelli, and P. Honeyman. Proposal: the Armored Packet
            Vault, September 2000. Draft.

[Say02]   O. Saydjari. LOCK: An Historical Perspective. In *18th Annual Computer Security Applications Conference*, pages 96–108, 2002. This was a "classic papers" retrospective of a paper from 1987.

[SCG⁺03a]   E. Suh, D. Clarke, G. Gassend, M. van Dijk, and S. Devadas. Efficient Memory Integrity Verification and Encryption for Secure Processors. In *Proceedings of the 36th Annual IEEE/ACM International Symposium on Microarchitecture (MICRO)*, pages 339–350, December 2003.

[SCG⁺03b]   G. Suh, D. Clarke, B. Gassend, M. van Dijk, and S. Devadas. AEGIS: Architecture for Tamper-Evident and Tamper-Resistant Processing. In *Proceedings of the 17th International Conference on Supercomputing*, pages 160–171, 2003.

[Sch02]   F. Schneider. Secure Systems Conundrum. *Communications of the ACM*, 45(10):160, October 2002.

[Sch03a]   S. Schoen. Trusted computing: Promise and risk. http://www.eff.org/Infra/trusted_computing/20031001_tc.php, October 2003.

[Sch03b]   S. Schoen. Who Controls Your Computer? Electronic Frontier Foundation Reports on Trusted Computing. http://www.eff.org/Infra/trusted_computing/20031002_eff_pr.php, October 2003.

[SCT04]   U. Shankar, M. Chew, and J.D. Tygar. Side Effects Are Not Sufficient to Authenticate Software. In *Proceedings of the 13th USENIX Security Symposium*, pages 89–102, 2004.

[Sec03]   Security Model for the Next-Generation Secure Computing Base. Windows Platform Design Notes, 2003.

[SGI]   SGI IRIX 6.5: Home Page. http://www.sgi.com/software/irix6.5.

[Shy]   sHype—Secure Hypervisor. http://www.research.ibm.com/secure_systems_department/projects/hypervisor/.

[SKv03]   D. Safford, J. Kravitz, and L. van Doorn. Take Control of TCPA. *Linux Journal*, pages 50–55, August 2003.

[Smi96]   S.W. Smith. Secure Coprocessing Applications and Research Issues. Technical Report Los Alamos Unclassified Release LA-UR-96-2805, Los Alamos National Laboratory, August 1996.

[Smi01]   S.W. Smith. WebALPS: A Survey of E-Commerce Privacy and Security Applications. *ACM SIGecom Exchanges*, 2.3:27–34, September 2001.

[Smi02]   S.W. Smith. Outbound Authentication for Programmable Secure Coprocessors. In *Computer Security—ESORICS 2002*, pages 72–89. Springer-Verlag LNCS 2502, October 2002. A revised and extended version will appear in the *International Journal of Information Security*.

[Smi03]   S.W. Smith. Fairy Dust, Secrets and the Real Worl. *IEEE Security and Privacy*, 1:89–93, January/February 2003.

[Smi04]   S.W. Smith. Probing End-User IT Security Practices—via Homework. *The Educause Quarterly*, 27, 2004. To appear.

[SPLK04]   A. Seshadri, A. Perrig, L.van Doorn, and P. Khosla. SWAtt: Software-based Attestation for Embedded Devices. In *Proceedings of the 2004 Symposium on Security and Privacy*, pages 272–282. IEEE, 2004.

[SPW98]   S.W. Smith, E. Palmer, and S. Weingart. Using a High-Performance, Programmable Secure Coprocessor. In *Financial Cryptography, Second International Conference, FC'98*, pages 73–89. Springer-Verlag LNCS 1465, 1998.

[SPWA99]   S.W. Smith, R. Perez, S.H. Weingart, and V. Austel. Validating a High-Performance, Programmable Secure Coprocessor. In *22nd National Information Systems Security Conference*, October 1999.

[SS01]   S.W. Smith and D. Safford. Practical Server Privacy Using Secure Coprocessors. *IBM Systems Journal*, 40:683–695, 2001.

[SS03]   A. Sadeghi and C Stuble. Taming "Trusted Platforms" by Operating System Design. In *Information Security Applications*, pages 286–302. Springer-Verlag LNCS 2908, 2003.

[SS04]   A. Sadeghi and C Stuble. Property-based Attestation for Computing Platforms: Caring about Properties, not Mechanisms. In *New Security Paradigms Workshop*, September 2004.

[ST98]   T. Sander and C. Tschudin. On Software Protection Via Function Hiding. In *2nd International Workshop on Information Hiding*, pages 111–123. Springer-Verlag LNCS 1525, 1998.

[ST04]   A. Shamir and E. Tramer. Acoustic cryptanalysis: On nosy people and noisy machines. Eurocrypt 2004 rump session, 2004. http://www.wisdom.weizmann.ac.il/~tromer/acoustic/.

[Sta03]   N. Stam. Inside Intel's Secretive 'LaGrande' Project. http://www.extremetech.com/, September 2003.

[Sun91]   SunOS SPARC Integer Division Vulnerability, 1991. CERT Advisory CA-91:16.

[SW99]   S.W. Smith and S. Weingart. Building a High-Performance, Programmable Secure Coprocessor. *Computer Networks*, 31:831–860, April 1999.

[SZJv04]   R. Sailer, X. Zhang, T. Jaeher, and L. van Doorn. Design and Implementation of a TCG-Based Integrity Measurement Architecture. In *Proceedings of the 13th USENIX Security Symposium*, pages 223–238, 2004.

[The03]   The Processor Resource/System Manager (PR/SM) for IBM zSeries z900 is awarded a certificate by the German Federal Office for Information Technology Security. Press release, March 2003.

[Tru01]   Trusted Computing Platform Alliance. TCPA PC Specific Implementation Specification, Version 1.00. http://www.trustedcomputinggroup.org, September 2001.

[Tru02]   Trusted Computing Platform Alliance. Main Specification, Version 1.1b. http://www.trustedcomputinggroup.org, February 2002.

[Tru03a]   Trusted Computing Group. TPM Main Part 1 Design Principles. `http://www.trustedcomputinggroup.org`, October 2003. Specification Version 1.2, Revision 62.

[Tru03b]   Trusted Computing Group. TPM Main Part 2 TPM Structures. `http://www.trustedcomputinggroup.org`, October 2003. Specification Version 1.2, Revision 62.

[Tru03c]   Trusted Computing Group. TPM Main Part 3 Commands. `http://www.trustedcomputinggroup.org`, October 2003. Specification Version 1.2, Revision 62.

[Tru04]   Trusted Computing Group. TCG Specification Architecture Overview. `http://www.trustedcomputinggroup.org`, April 2004. Specification Revision 1.2.

[TY91]   J.D. Tygar and B.S. Yee. Strongbox: A System for Self-Securing Programs. In *CMU Computer Science: A 25th Anniversary Commemorative*, pages 163–197. Addison-Wesley, 1991.

[TY93]   J.D. Tygar and B.S. Yee. Dyad: A System for Using Physically Secure Coprocessors. In *Proceedings of the Joint Harvard-MIT Workshop on Technological Strategies for the Protection of Intellectual Property in the Network Multimedia Environment*, April 1993.

[TYH96]   J. D. Tygar, B. S. Yee, and N. Heintze. cryptographic Postage Indicia. In *Asian Computing Science Conference*, pages 378–391, 1996.

[van85]   W. van Eck. Electromagnetic Radiation from Video Display Units: An Eavesdropping Risk? *Computers and Security*, 4:269–286, 1985.

[van03]   P.C. van Oorschot. Revisiting Software Protection. In *Information Security, 6th International Conference, ISC 2003*, pages 1–13. Springer-Verlag LNCS 2851, 2003.

[vGA01]   L. van Doorn, G.Ballintijn, and W. Arbaugh. Signed Executables for Linux. Technical Report UMD CS-TR-4259, University of Maryland, June 2001.

[VS04]   G. Vanrenen and S.W. Smith. Distributing Security-Mediated PKI. In *1st European PKI Workshop: Research and Applications*, pages 218–231. Springer-Verlag LNCS 3093, 2004.

[Wak68]   A. Waksman. A Permutation Network. *Journal of the ACM*, 15(1):159–163, 1968.

[WC87]   S.R. White and L.D. Comerford. ABYSS: A Trusted Architecture for Software Protection. In *IEEE Symposium on Security and Privacy*, 1987.

[Wei87]   S.H. Weingart. Physical Security for the μABYSS System. In *Proceedings of the 1987 Symposium on Security and Privacy*, pages 52–59. IEEE, 1987.

[Wei00]   S. Weingart. Physical Security Devices for Computer Subsystems: A Survey of Attacks and Defenses. In *Cryptographic Hardware and Embedded Systems— CHES 2000*, pages 302–317. Springer-Verlag LNCS 1965, 2000.

[WWAD90]   S.H. Weingart, S. White, W. Arnold, and G. Double. An Evaluation System for the Physical Security of Computing Systems. In *Sixth Annual Computer Security Applications Conference*, pages 232–243, 1990.

[WWAP91]   S. White, S.H. Weingart, W. Arnold, and E. R. Palmer. Introduction to the Citadel Architecture: Security in Physically Exposed Environments. Technical Report RC16672, IBM T.J. Watson Research Center, 1991.

[Yee94]   B.S. Yee. *Using Secure Coprocessors*. PhD thesis, Carnegie Mellon University, May 1994. Page numbers refer to the LATEX 2E version.

[Yee99]   B.S. Yee. A Sanctuary for Mobile Agents. In *Secure Internet Programming: Security Issues for Mobile and Distributed Objects*, pages 261–274. Springer-Verlag LNCS 1603, 1999.

[YS02]   E. Ye and S.W. Smith. Trusted Paths for Browsers. In *Proceedings of the 11th USENIX Security Symposium*, August 2002.

[YT95]   B.S. Yee and J.D. Tygar. Secure Coprocessors in Electronic Commerce Applications. In *Proeedings of the 1st USENIX Electronic Commerce Workshop*, pages 155–170. USENIX, 1995.

[YY96]   A. Young and M. Yung. The Dark Side of Black-Box Cryptography, or: Should We Trust Capstone? In *Advances in Cryptology—Crypto 96*, pages 89–103. Springer-Verlag LNCS 1109, 1996.

[YYS02]   E. Ye, Y. Yuan, and S.W. Smith. Web Spoofing Revisited: SSL and Beyond. Technical Report TR2002-417, Department of Computer Science, Dartmouth College., 2002.

[ZZPL03]   X. Zhuang, T. Zhang, S. Pande, and H.H.S. Lee. HIDE: Hardware Support for Leakage-Immune Dynamic Execution. Technical Report GIT-CERCS-03-21, Georgia Institute of Technology, 2003.

About the Author

 **Sean Smith** is currently on the faculty of the Department of Computer Science at Dartmouth College, serves as director of the Cyber Security and Trust Research Center at Dartmouth's Institute for Security Technology Studies, and also serves as Principal Investigator of the Dartmouth PKI Lab. His current research and teaching focus on how to build trustworthy systems in the real world. He previously worked as a scientist at IBM T.J. Watson Research Center, doing secure coprocessor design, implementation and validation; and at Los Alamos National Laboratory, doing security designs and analyses for a wide range of public-sector clients. Dr. Smith was educated at Princeton (B.A., Math; rugby team) and Carnegie Mellon (M.S., Ph.D., Computer Science; cycling team). If you corner him at a conference, he might start talking about trail-running, orienteering, ultra-marathons, or bicycles.

http://www.cs.dartmouth.edu/~sws/

Index